Gene Expression Profiling

METHODS IN MOLECULAR BIOLOGY™

John M. Walker, SERIES EDITOR

METHODS IN MOLECULAR BIOLOGY™

Gene Expression Profiling

Methods and Protocols

Edited by

Richard A. Shimkets

CuraGen Corporation, New Haven, CT

HUMANA PRESS ✳ TOTOWA, NEW JERSEY

© 2004 Humana Press Inc.
999 Riverview Drive, Suite 208
Totowa, New Jersey 07512

www.humanapress.com

This publication is printed on acid-free paper. ∞
ANSI Z39.48-1984 (American Standards Institute)

Permanence of Paper for Printed Library Materials.
Cover illustration: From Fig. 1, in Chapter 4, "GeneCalling: Transcript Profiling Coupled to a Gene Database Query," by Richard A. Shimkets.

Production Editor: Robin B. Weisberg
Cover design by Patricia F. Cleary.

For additional copies, pricing for bulk purchases, and/or information about other Humana titles, contact Humana at the above address or at any of the following numbers: Tel.: 973-256-1699; Fax: 973-256-8341; E-mail: humana@humanapr.com; or visit our Website: www.humanapress.com

Printed in the United States of America. 10 9 8 7 6 5 4 3 2 1

E-ISBN 1-59259-751-3

Library of Congress Cataloging-in-Publication Data

Gene expression profiling : methods and protocols / edited by Richard A. Shimkets.
    p. ; cm. -- (Methods in molecular biology, ISSN 1064-3745 ; v.258)
Includes bibliographical references and index.
  ISBN 1-58829-220-7 (alk. paper)
  1. Gene expression--Laboratory manuals.
    [DNLM: 1. Gene Expression Profiling--methods. QZ 52 G326 2004] I. Shimkets, Richard A.
II. Series: Methods in molecular biology (Totowa, N.J.) v. 258.
  QH450.G46255 2004
  572.8'65--dc22
                                                              2003022875

Preface

Why Quantitate Gene Expression?

The central dogma of molecular biology is the concept of the transcription of messenger RNA from a DNA template and translation of that RNA into protein. Since the transcription of RNA is a key regulatory point that may eventually signal many other cascades of events, the study of RNA levels in a cell or organ can help the understanding of a wide variety of biological systems.

It is assumed that readers of *Gene Expression Profiling: Methods and Protocols* have an appreciation for this fact, so that the present volume focuses on the practical and technical considerations that guide the choice of methodology in this area. Thoroughly thinking through the specific scientific questions at hand will enable the choice of the best technology for that application and examples of which technologies fit which applications best will be discussed.

Richard A. Shimkets

Contents

Contributors

LUANNE CHEHAK • *Third Wave Technologies Inc., Madison, WI*

ZHIJIAN J. CHEN • *Department of Pharmacology, Fox Chase Cancer Center, Philadelphia, PA*

MARINA E. CHICUREL • *Affymetrix Inc., Santa Clara, CA*

ERIN L. CRAWFORD • *Departments of Medicine and Pathology, Medical College of Ohio, Toledo, OH*

MICHELLE L. CURTIS • *Third Wave Technologies Inc., Madison, WI*

DENNISE D. DALMA-WEISZHAUSZ • *Affymetrix Inc., Santa Clara, CA*

SEJAL M. DESAI • *BD Biosciences Clontech, Palo Alto, CA*

TIMOTHY G. GRAVES • *Departments of Medicine and Pathology, Medical College of Ohio, Toledo, OH*

ELIZABETH A. HERNESS • *Gene Express Inc., Toledo, OH*

CHARLES R. KNIGHT • *Departments of Medicine and Pathology, Medical College of Ohio, Toledo, OH*

ROBERT W. KWIATKOWSKI • *Third Wave Technologies Inc., Madison, WI*

SERGEY A. LUKYANOV • *Evrogen JSC and Institute of Bioorganic Chemistry, Russian Academy of Sciences, Moscow, Russia*

MARTIN D. LEACH • *Bioinformatics, CuraGen Corporation, New Haven, CT*

JENNIFER LESCALLETT • *Affymetrix,Inc., Santa Clara, CA*

ROBERT LIPSHUTZ • *Affymetrix Inc., Santa Clara, CA*

CHERYL A. MOTTEN • *Departments of Medicine and Pathology, Medical College of Ohio, Toledo, OH*

MARILYN C. OLSON • *Third Wave Technologies Inc., Madison, WI*

SARAH M. OLSON • *Third Wave Technologies Inc., Madison, WI*

DENIS V. REBRIKOV • *Evrogen JSC and Institute of Bioorganic Chemistry, Russian Academy of Sciences, Moscow, Russia*

RICHARD A. SHIMKETS • *Drug Discovery, CuraGen Corporation, New Haven, CT*

PAUL D. SIEBERT • *BD Biosciences Clontech, Palo Alto, CA*

TSETSKA TAKOVA • *Third Wave Technologies Inc., Madison, WI*

KENNETH D. TEW • *Department of Pharmacology, Fox Chase Cancer Center, Philadelphia, PA*

GILBERT VASSART • *Institute of Interdisciplinary Research (IRIBHM) Université Libre de Bruxelles, Brussels, Belgium and Department of Medical Genetics of the Erasme Hospital, Brussels, Belgium*

CATHELINE VILAIN • *Institute of Interdisciplinary Research (IRIBHM) Université Libre de Bruxelles, Brussels, Belgium*
K. A. WARNER • *Departments of Medicine and Pathology, Medical College of Ohio, Toledo, OH*
JAMES C. WILLEY • *Departments of Medicine and Pathology, Medical College of Ohio, Toledo, OH*
ROBERT J. ZAHORCHAK • *Gene Express Inc., Toledo, OH*

1

Technical Considerations in Quantitating Gene Expression

Richard A. Shimkets

1. Introduction

Scientists routinely lecture and write about gene expression and the abundance of transcripts, but in reality, they extrapolate this information from a variety of measurements that different technologies may provide. Indeed, there are many reasons that applying different technologies to transcript abundance may give different results. This may result from an incomplete understanding of the gene in question or from shortcomings in the applications of the technologies.

The first key factor to appreciate in measuring gene expression is the way that genes are organized and how this influences the transcripts in a cell. **Figure 1** depicts some of the scenarios that have been determined from sequence analyses of the human genome. Most genes are composed of multiple exons transcribed with intron sequences and then spliced together. Some genes exist entirely between the exons of other genes, either in the forward or reverse orientation. This poses a problem because it is possible to recover a fragment or clone that could belong to multiple genes, be derived from an unspliced transcript, or be the result of genomic DNA contaminating the RNA preparation. All of these events can create confusing and confounding results. Additionally, the gene duplication events that have occurred in organisms that are more complex have led to the existence of closely related gene families that coincidentally may lie near each other in the genome. In addition, although there are probably less than 50,000 human genes, the exons within those genes can be spliced together in a variety of ways, with some genes documented to produce more than 100 different transcripts *(1)*.

From: *Methods in Molecular Biology, Vol. 258: Gene Expression Profiling: Methods and Protocols*
Edited by: R. A. Shimkets © Humana Press Inc., Totowa, NJ

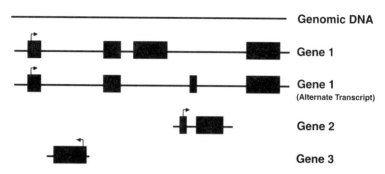

Fig. 1. Typical gene exon structure.

Therefore, there may be several hundred thousand distinct transcripts, with potentially many common sequences. Gene biology is even more interesting and complex, however, in that genetic variations in the form of single nucleotide polymorphisms (SNPs) frequently cause humans and diploid or polyploid model systems to have two (or more) distinct versions of the same transcript.

This set of facts negates the possibility that a single, simple technology can accurately measure the abundance of a specific transcript. Most technologies probe for the presence of pieces of a transcript that can be confounded by closely related genes, overlapping genes, incomplete splicing, alternative splicing, genomic DNA contamination, and genetic polymorphisms. Thus, independent methods that verify the results in different ways to the exclusion of confounding variables are necessary, but frequently not employed, to gain a clear understanding of the expression data. The specific means to work around these confounding variables are mentioned here, but a blend of techniques will be necessary to achieve success.

2. Methods and Considerations

There are nine basic considerations for choosing a technology for quantitating gene expression: architecture, specificity, sensitivity, sample requirement, coverage, throughput, cost, reproducibility, and data management.

2.1. Architecture

We define the architecture of a gene-expression analysis system as either an open system, in which it is possible to discover novel genes, or a closed system in which only known gene or genes are queried. Depending on the application, there are numerous advantages to open systems. For example, an open system may detect a relevant biological event that affects splicing or genetic variation. In addition, the most innovative biological discovery processes have involved the

discovery of novel genes. However, in an era where multiple genome sequences have been identified, this may not be the case. The genomic sequence of an orga-   ⁰⁾
nism, however, has not proven sufficient for the determination of all of the transcripts encoded by that genome, and thus there remain prospects for novelty regardless of the biological system. In model systems that are relatively uncharacterized at the genomic or transcript level, entire technology platforms may be excluded as possibilities. For example, if one is studying transcript levels in a rabbit, one cannot comprehensively apply a hybridization technology because there are not enough transcripts known for this to be of value. If one simply wants to know the levels of a set of known genes in an organism, a hybridization technology may be the most cost-effective, if the number of genes is sufficient to warrant the cost of producing a gene array.

2.2. Specificity

The evolution of genomes through gene or chromosomal fragment duplications and the subsequent selection for their retention, has resulted in many gene families, some of which share substantial conservation at the protein and nucleotide level. The ability for a technology to discriminate between closely related gene sequences must be evaluated in this context in order to determine whether one is measuring the level of a single transcript, or the combined, added levels of multiple transcripts detected by the same probing means. This is a double-edged sword because technologies with high specificity, may fail to identify one allele, or may do so to a different degree than another allele when confronted with a genetic polymorphism. This can lead to the false positive of an expression differential, or the false negative of any expression at all. This is addressed in many methods by surveying multiple samples of the same class, and probing multiple points on the same gene. Methods that do this effectively are preferred to those that do not.

2.3. Sensitivity

The ability to detect low-abundance transcripts is an integral part of gene discovery programs. Low-abundance transcripts, in principle, have properties that are of particular importance to the study of complex organisms. Rare transcripts frequently encode for proteins of low physiologic concentrations that in many cases make them potent by their very nature. Erythropoietin is a classic example of such a rare transcript. Amgen scientists functionally cloned erythropoietin long before it appeared in the public expressed sequence tag (EST) database. Genes are frequently discovered in the order of transcript abundance, and a simple analysis of EST databases correctly reveals high, medium, and low abundance transcripts by a direct correlation of the number of occurrences in that database (data not shown). Thus, using a technology that is more sensitive has the potential to identify novel transcripts even in a well-studied system.

Sensitivity values are quoted in publications for available technologies at concentrations of 1 part in 50,000 to 1 part in 500,000. The interpretation of these data, however, should be made cautiously both upon examination of the method in which the sensitivity was determined, as well as the sensitivity needed for the intended use. For example, if one intends to study appetite-signaling factors and uses an entire rat brain for expression analysis, the dilution of the target cells of anywhere from 1 part in 10,000 to 1 part in 100,000 allows for only the most abundant transcripts in the rare cells to be measured, even with the most sensitive technology available. Reliance on cell models to do the same type of analysis, where possible, suffers the confounding variable that isolated cells or cell lines may respond differently in culture at the level of gene expression. An ideal scenario would be to carefully micro dissect or sort the cells of interest and study them directly, provided enough samples can be obtained.

In addition to the ability of a technology to measure rare transcripts, the sensitivity to discern small differentials between transcripts must be considered. The differential sensitivity limit has been reported for a variety of techniques ranging from 1.5-fold to 5-fold, so the user must determine how important small modulations are to the overall project and choose the technology while taking this property into account as well.

2.4. Sample Requirement

The requirement for studying transcript abundance levels is a cell or tissue substrate, and the amount of such material needed for analysis can be prohibitively high with many technologies in many model systems. To use the above example, dozens of dissected rat hypothalami may be required to perform a global gene expression study, depending on the quantitating technology chosen. Samples procured by laser-capture microdissection can only be used in the measuring of a small number of transcripts and only with some technologies, or must be subjected to amplification technologies, which risk artificially altering transcript ratios.

2.5. Coverage

For open architecture systems where the objective is to profile as many transcripts as possible and identify new genes, the number of independent transcripts being measured is an important metric. However, this is one of the most difficult parameters to measure, because determining what fraction of unknown transcripts is missing is not possible. Despite this difficulty, predictive models can be made to suggest coverage, and the intuitive understanding of the technology is a good gage for the relevance and accuracy of the predictive model.

The problem of incomplete coverage is perhaps one of the most embarrassing examples of why hundreds of scientific publications were produced in the 1970's and 1980's having relatively little value. Many of these papers reported the identification of a single differentially expressed gene in some model system and expounded upon the overwhelmingly important new biological pathway uncovered. Modern analysis has demonstrated that even in the most similar biological systems or states, finding 1% of transcripts with differences is common, with this number increasing to 20% of transcripts or more for systems when major changes in growth or activation state are signaled. In fact, the activation of a single transcription factor can induce the expression of hundreds of genes. Any given abundantly altered transcript without an understanding of what other transcripts are altered, is similar to independent observers describing the small part of an elephant that they can see. The person looking at the trunk describes the elephant as long and thin, the person observing an ear believes it to be flat, soft and furry, and the observer examining a foot describes the elephant as hard and wrinkly. Seeing the list of the majority of transcripts that are altered in a system is like looking at the entire elephant, and only then can it be accurately described. Separating the key regulatory genes on a gene list from the irrelevant changes remains one of the biggest challenges in the use of transcript profiling.

2.6. Throughput

The throughput of the technology, as defined by the number of transcript samples measured per unit time, is an important consideration for some projects. When quick turnaround is desired, it is impractical to print microarrays, but where large numbers of data points need to be generated, techniques where individual reactions are required are impractical. Where large experiments on new models generate significant expense, it may be practical to perform a higher throughput, lower quality assay as a control prior to a large investment. For example, prior to conducting a comprehensive gene profiling experiment in a drug dose-response model, it might be practical to first use a low throughput technique to determine the relevance of the samples prior to making the investment with the more comprehensive analysis.

2.7. Cost

Cost can be an important driver in the decision of which technologies to employ. For some methods, substantial capital investment is required to obtain the equipment needed to generate the data. Thus, one must determine whether a microarray scanner or a capillary electrophoresis machine is obtainable, or if X-ray film and a developer need to suffice. It should be noted that as large companies change platforms, used equipment becomes available at prices dramatically less than those for brand new models. In some cases, homemade equipment can serve the purpose as well as commercial apparatuses at a fraction of the price.

2.8. Reproducibility

It is desired to produce consistent data that can be trusted, but there is more value to highly reproducible data than merely the ability to feel confident about the conclusions one draws from them. The ability to forward-integrate the findings of a project and to compare results achieved today with results achieved next year and last year, without having to repeat the experiments, is key to managing large projects successfully. Changing transcript-profiling technologies often results in datasets that are not directly comparable, so deciding upon and persevering with a particular technology has great value to the analysis of data in aggregate. An excellent example of this is with the serial analysis of gene expression (SAGE) technique, where directly comparable data have been generated by many investigators over the course of decades and are available online (http://www.ncbi.nlm.nih.gov).

2.9. Data Management

Management and analysis of data is the natural continuation to the discussion of reproducibility and integration. Some techniques, like differential display, produce complex data sets that are neither reproducible enough for subsequent comparisons, nor easily digitized. Microarray and GeneCalling data, however, can be obtained with software packages that determine the statistical significance of the findings and even can organize the findings by molecular function or biochemical pathways. Such tools offer a substantial advance in the generation of accretive data. The field of bioinformatics is flourishing as the number of data points generated by high throughput technologies has rapidly exceeded the number of biologists to analyze the data.

Reference

1. Ushkaryov, Y. A. and Sudhof, T. C. (1993) Neurexin IIIα: extensive alternative splicing generates membrane-bound and soluble forms. *Proc. Natl. Acad. Sci. USA* **90**, 6410–6414.

2

Gene Expression Quantitation Technology Summary

Richard A. Shimkets

Summary

Scientists routinely talk and write about gene expression and the abundance of transcripts, but in reality they extrapolate this information from the various measurements that a variety of different technologies provide. Indeed, there are many reasons why applying different technologies to the problem of transcript abundance may give different results, owing to an incomplete understanding of the gene in question or from shortcomings in the applications of the technologies. There are nine basic considerations for making a technology choice for quantitating gene expression that will impact the overall outcome: architecture, specificity, sensitivity, sample requirement, coverage, throughput, cost, reproducibility, and data management. These considerations will be discussed in the context of available technologies.

Key Words: Architecture, bioinformatics, coverage, quantitative, reproducibility, sensitivity, specificity, throughput

1. Introduction

Owing to the intense interest of many groups in determining transcript levels in a variety of biological systems, there are a large number of methods that have been described for gene-expression profiling. Although the actual catalog of all techniques developed is quite extensive, there are many variations on similar themes, and thus we have reduced what we present here to those techniques that represent a distinct technical concept. Within these groups, we discovered that there are methods that are no longer applied in the scientific community, not even in the inventor's laboratory. Thus, we have chosen to focus the methods chapters of this volume on techniques that are in common use in the community

From: *Methods in Molecular Biology, Vol. 258: Gene Expression Profiling: Methods and Protocols*
Edited by: R. A. Shimkets © Humana Press Inc., Totowa, NJ

at the time of this writing. This work also introduces two novel technologies, SEM-PCR and the Invader Assay, that have not been described previously. Although these methods have not yet been formally peer-reviewed by the scientific community, we feel these approaches merit serious consideration.

In general, methods for determining transcript levels can be based on transcript visualization, transcript hybridization, or transcript sequencing **(Table 1)**.

The principle of transcript visualization methods is to generate transcripts with some visible label, such as radioactivity or fluorescent dyes, to separate the different transcripts present, and then to quantify by virtue of the label the relative amount of each transcript present. Real-time methods for measuring label while a transcript is in the process of being linearly amplified offer an advantage in some cases over methods where a single time-point is measured. Many of these methods employ the polymerase chain reaction (PCR), which is an effective way of increasing copies of rare transcripts and thus making the techniques more sensitive than those without amplification steps. The risk to any amplification step, however, is the introduction of amplification biases that occur when different primer sets are used or when different sequences are amplified. For example, two different genes amplified with gene-specific primer sets in adjacent reactions may be at the same abundance level, but because of a thermodynamic advantage of one primer set over the other, one of the genes might give a more robust signal. This property is a challenge to control, except by multiple independent measurements of the same gene. In addition, two allelic variants of the same gene may amplify differently if the polymorphism affects the secondary structure of the amplified fragment, and thus an incorrect result may be achieved by the genetic variation in the system. As one can imagine, transcript visualization methods do not provide an absolute quantity of transcripts per cell, but are most useful in comparing transcript abundance among multiple states.

Transcript hybridization methods have a different set of advantages and disadvantages. Most hybridization methods utilize a solid substrate, such as a microarray, on which DNA sequences are immobilized and then labeled. Test DNA or RNA is annealed to the solid support and the locations and intensities on the solid support are measured. In another embodiment, transcripts present in two samples at the same levels are removed in solution, and only those present at differential levels are recovered. This suppression subtractive hybridization method can identify novel genes, unlike hybridizing to a solid support where information generated is limited to the gene sequences placed on the array. Limitations to hybridization are those of specificity and sensitivity. In addition, the position of the probe sequence, typically 20–60 nucleotides in length, is critical to the detection of a single or multiple splice variants. Hybridization methods employing cDNA libraries instead of synthetic oligonucleotides give inconsistent results, such as variations in splicing and not allowing for the testing of the levels of putative transcripts predicted from genomic DNA sequence.

Hybridization specificity can be addressed directly when the genome sequence of the organism is known, because oligonucleotides can be designed specifically to detect a single gene and to exclude the detection of related genes. In the absence of this information, the oligonucleotides cannot be designed to assure specificity, but there are some guidelines that lead to success. Protein-coding regions are more conserved at the nucleotide level than untranslated regions, so avoiding translated regions in favor of regions less likely to be conserved is useful. However, a substantial amount of alternative splicing occurs immediately distal to the 3' untranslated region and thus designing in proximity to regions following the termination codon may be ideal in many cases. Regions containing repetitive elements, which may occur in the untranslated regions of transcripts, should be avoided.

Several issues make the measurement of transcript levels by hybridization a relative measurement and not an absolute measurement. Those experienced with hybridization reactions recognize the different properties of sequences annealing to their complementary sequences, and thus empirical optimization of temperatures and wash conditions have been integrated into these methods.

Principle disadvantages to hybridization methods, in addition to those of any closed system, center around the analysis of what is actually being measured. Typically, small regions are probed and if an oligonucleotide is designed to a region that is common to multiple transcripts or splice variants, the resulting intensity values may be misleading. If the oligonucleotide is designed to an exon that is not used in one sample of a comparison, the results will indicate lack of expression, which is incorrect. In addition, hybridization methods may be less sensitive and may yield a negative result when a positive result is clearly present through visualization.

The final class of technologies that measure transcript levels, transcript sequencing, and counting methods can provide absolute levels of a transcript in a cell. These methods involve capturing the identical piece of all genes of interest, typically the 3' end of the transcript, and sequencing a small piece. The number of times each piece was sequenced can be a direct measurement of the abundance of that transcript in that sample. In addition to absolute measurement, other principle advantages of this method include the simplicity of data integration and analysis and a general lack of problems with similar or overlapping transcripts. Principle disadvantages include time and cost, as well as the fact that determining the identity of a novel gene by only the 10-nucleotide tag is not trivial.

We would like to mention two additional considerations before providing detailed descriptions of the most popular techniques. The first is contamination

Table 1
Common Gene Expression Profiling Methods

Technique	Class	Architecture	Kits Available	Service Available	Detect Alt. Splicing	Detect SNPs
5'-nuclease assay/real-time RT-PCR	Visualization	Open	Yes	No	No	No
AFLP (amplified-fragment length polymorphism fingerprinting)	Visualization	Open	No	No	No	Yes
Antisense display	Visualization	Open	No	No	No	No
DDRT-PCR (differential display RT-PCR)	Visualization	Open	Yes	No	No	No
DEPD (digital expression pattern display)	Visualization	Open	No	No	Yes	No
Differential hybridization (differential cDNA library screening)	Hybridization	Open	No	No	No	No
DSC (differential subtraction chain)	Hybridization	Open	No	No	No	No
GeneCalling	Visualization	Open	No	Yes	Yes	Yes
In situ Hybridization	Hybridization	Closed	Yes	No	No	No
Invader Assay	Visualization	Closed	Yes	Yes	No	Yes
Microarray hybridization	Hybridization	Closed	Yes	Yes	No	No
Molecular indexing (and computational methods)	Visualization	Open	No	No	No	No
MPSS (massively parallel signature sequencing)	Sequencing	Open	No	No	No	No
Northern-Blotting (Dot-/Slot-Blotting)	Hybridization	Closed	Yes	No	No	No
Nuclear run on assay/nuclease S1 analysis	Visualization	Closed	Yes	No	No	No
ODD (ordered differential display)	Visualization	Open	No	No	No	No
Quantitative RT-PCR	Visualization	Closed	Yes	Yes	No	No

RAGE (rapid analysis of gene expression)	Visualization	Open	No	No	Yes	No
RAP-PCR (RNA arbitrarily primed PCR fingerprinting)	Visualization	Open	No	No	No	No
RDA (representational difference analysis)	Visualization	Open	No	No	No	No
RLCS (restriction landmark cDNA scanning)	Visualization	Open	No	No	No	No
RPA (ribonuclease protection assay)	Visualization	Open	No	No	No	No
RSDD (reciprocal subtraction differential display)	Visualization	Open	No	No	No	No
SAGE (serial analysis of gene expression)	Sequencing	Open	Yes	No	No	No
SEM-PCR	Visualization	Closed	No	Yes	No	No
SSH (suppression subtractive hybridization)	Hybridization	Open	Yes	No	Yes	No
Suspension arrays with microbeads	Hybridization	Closed	No	No	No	No
TALEST (tandem arrayed ligation of expressed sequence tags)	Sequencing	Open	No	No	No	No

of genomic or mitochondrial DNA or unspliced RNA contamination in messenger RNA preparations. Even using oligo-dT selection and DNAse digestion, DNA and unspliced RNA tends to persist in many RNA preparations. This is evidenced by an analysis of the human expressed sequence tag (EST) database for sequences obtained that are clearly intronic or intragenic. These sequences tile the genome evenly and comprise from 0.5% to up to 5% of the ESTs in a given sequencing project, across even the most experienced sequencing centers (unpublished observation). Extremely sensitive technologies can detect the contaminating genomic DNA and give false-positive results. A common mistake when using quantitative PCR methods involves the use of gene-specific primers to design the primers within the same exon. This often yields a positive result because a few copies of genomic DNA targets will be present. By designing primer sets that span large introns, a positive result excludes both genomic DNA contamination as well as unspliced transcripts. This is not always possible, of course, in the cases of single-exon genes like olfactory G protein-coupled receptors and in organisms like saccharomyces and fungi where multi-exon genes are not common. In these cases, a control primer set that will only amplify genomic DNA can aid dramatically in the interpretation of the results.

A final, and practical consideration is to envision the completion of the project of interest, because using different quantitation methods will result in the need for different follow-up work. For example, if a transcript counting method that reveals 10 nucleotides of sequence is used, how will those data be followed up? What prioritization criteria for the analysis will be used, and how will the full-length sequences and full-length clones, for those genes be obtained? This may sound like a trivial concern, but in actuality, the generation of large sets of transcript-abundance data may create a quantity of follow-up work that may be unwieldy or even unreasonable. Techniques that capture the protein-coding regions of transcripts, such as GeneCalling, reveal enough information for many novel genes that may help prioritize their follow-up, rather than 3'-based methods where there is little ability to prioritize follow-up without a larger effort. Beginning with the completion of the project in mind allows the researcher to maximize the time line and probability for completion, as well as produce the best quality research result in the study of gene expression.

3

Standardized RT-PCR and the Standardized Expression Measurement Center

James C. Willey, Erin L. Crawford, Charles R. Knight,
K. A. Warner, Cheryl A. Motten, Elizabeth A. Herness,
Robert J. Zahorchak, and Timothy G. Graves

Summary

Standardized reverse transcriptase polymerase chain reaction (StaRT-PCR) is a modification of the competitive template (CT) RT method described by Gilliland et al. StaRT-PCR allows rapid, reproducible, standardized, quantitative measurement of data for many genes simultaneously. An internal standard CT is prepared for each gene, cloned to generate enough for $>10^9$ assays and CTs for up to 1000 genes are mixed together. Each target gene is normalized to a reference gene to control for cDNA loaded in a standardized mixture of internal standards (SMIS) into the reaction. Each target gene and reference gene is measured relative to its respective internal standard within the SMIS. Because each target gene and reference gene is simultaneously measured relative to a known number of internal standard molecules in the SMIS, it is possible to report each gene expression measurement as a numerical value in units of target gene cDNA molecules/ 10^6 reference gene cDNA molecules. Calculation of data in this format allows for entry into a common databank, direct interexperimental comparison, and combination of values into interactive gene expression indices.

Key Words: cDNA, expression, mRNA, quantitative, RT- PCR, StaRT-PCR

1. Introduction

With the recent completion of the human genome project, attention is now focusing on functional genomics. In this context, a key task is to understand normal and pathological function by empirically correlating gene expression patterns with known and newly discovered phenotypes. As with other areas of science, progress in this area will accelerate greatly when there is an accepted standardized way to measure gene expression *(1,2)*.

From: *Methods in Molecular Biology, Vol. 258: Gene Expression Profiling: Methods and Protocols*
Edited by: R. A. Shimkets © Humana Press Inc., Totowa, NJ

Standardized reverse transcriptase-polymerase chain reaction (StaRT-PCR) is a modification of the competitive template (CT) reverse transcriptase (RT) method described by Gilliland et al. *(3)*. StaRT-PCR allows rapid, reproducible, standardized, and quantitative measurement of data for many genes simultaneously *(4–15)*. An internal standard CT is prepared for each target gene and reference gene (e.g., β-actin and GAPDH), then cloned to generate enough for >10^9 assays. Internal standards for up to 1000 genes are quantified and mixed together in a standardized mixture of internal standards (SMIS). Each target gene is normalized to a reference gene to control for cDNA loaded into the reaction. Each target gene and reference gene is measured relative to its respective internal standard in the SMIS. Because each target gene and reference gene is simultaneously measured relative to a known number of internal standard molecules that have been combined into the SMIS, it is possible to report each gene expression measurement as a numerical value in units of target gene cDNA molecules/ 10^6 reference gene cDNA molecules. Calculation of data in this format allows for entry into a common databank *(5)*, direct interexperimental comparison *(4–15)*, and combination of values into interactive gene expression indices *(8,9,11)*.

With StaRT-PCR, as is clear in the schematic presented in **Fig. 1A**, expression of each reference gene (e.g., β-actin) or target gene (e.g., Gene 1–6) in a sample (for example sample A) is measured relative to its respective internal standard in the SMIS. Because in each experiment the internal standard for each gene is present at a fixed concentration relative to all other internal standards, it is possible to quantify the expression of each gene relative to all others measured. Furthermore, it is possible to compare data from analysis of sample A to those from analysis of all other samples, represented as B_{1-n}. This result is a continuously expanding virtual multiplex experiment. That is, data from an ever-expanding number of genes and samples may be entered into the same database. Because the number of molecules for each standard is known, it is possible to calculate all data in the form of molecules/reference gene molecules.

In contrast, for other multigene methods, such as multiplex real-time RT-PCR or microarrays, represented in **Fig. 1B**, expression of each gene is directly compared from one sample to another and data are in the form of fold differences. Because of intergene variation in hybridization efficiency and/or PCR amplification efficiency, and the absence of internal standards to control for these sources of variation, it is not possible to directly compare expression of one gene to another in a sample or to obtain values in terms of molecules/molecules of reference gene.

In numerous studies, StaRT-PCR has provided both intralaboratory *(4–15)* and interlaboratory reproducibility *(6)* sufficient reproducability to detect two-fold differences in gene expression. StaRT-PCR identifies interactive gene expression indices associated with lung cancer *(8–10)*, pulmonary sarcoidosis

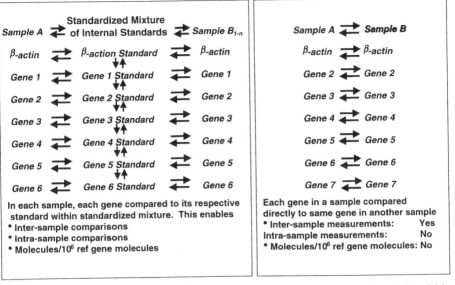

Fig. 1 **(A)** Schematic diagram of the relationship among internal standards within the SMIS and between each internal standard and its respective cDNA from a sample. The internal standard for each reference gene and target gene is at a fixed concentration relative to all other internal standards within the SMIS. Within a polymerase chain reaction (PCR) master mixture, in which a cDNA sample is combined with SMIS, the concentration of each internal standard is fixed relative to the cDNA representing its respective gene. In the PCR product from each sample, the number of cDNA molecules representing a gene is measured relative to its respective internal standard rather than by comparing it to another sample. Because everyone uses the same SMIS, and there is enough to last 1000 years at the present rate of consumption, all gene expression measurements may be entered into the same database. **(B)** Measurement by multiplex RT-PCR or microarray analysis. Using these methods each gene scales differently because of gene-to-gene variation in melting temperature between gene and PCR primers or gene and sequence on microarray. Consequently, it is possible to compare relative differences in expression of a gene from one sample to another, but not difference in expression among many genes in a sample. Further, it is not possible to develop a reference database, except in relationship to a nonrenewable calibrator sample. Moreover, unless a known quantity of standard template is prepared for each gene, it is not possible to know how many copies of a gene are expressed in the calibrator sample, or the samples that are compared to the calibrator.

(13), cystic fibrosis *(14)*, and chemoresistance in childhood leukemias *(11)*. In a recent report, StaRT-PCR methods provided reproducible gene expression measurement when StaRT-PCR products were separated and analyzed by matrix-

assisted laser desorption/ionization-time of flight mass spectrometry (MALDI-TOF MS) instead of by electrophoresis *(16)*.

In a recent multi-institutional study *(6)*, data generated by StaRT-PCR were sufficiently reproducible to support development of a meaningful gene expression database and thereby serve as a common language for gene expression.

StaRT-PCR is easily adapted to automated systems and readily subjected to quality control. Recently, we established the National Cancer Institute-funded Standardized Expression Measurement (SEM) Center at the Medical College of Ohio that utilizes robotic systems to conduct high-throughput StaRT-PCR gene expression measurement. In the SEM Center, the coefficient of variance (CV) for StaRT-PCR is less than 15%.

In this chapter, we describe in detail the StaRT-PCR method, comparing and contrasting StaRT-PCR to real-time RT-PCR, a well-established quantitative RT-PCR method. In addition, we describe the SEM Center, including the equipment and methods used, how to access it, and the type of data produced.

2. StaRT-PCR vs Real-Time RT-PCR

There are several potential sources of variation in quantitative RT-PCR gene expression measurement, as outlined in **Table 1**.

StaRT-PCR, by including internal standards in the form of a SMIS in each gene expression measurement, controls for each of these sources of variation. In contrast, using real-time RT-PCR without internal standards, it is possible to control for some, but not all of these sources of variation. Additionally, with real-time RT-PCR, control often requires external standard curves, and these add time and are themselves a potential source of error. These issues are discussed in this section.

2.1. Control for Variation in Loading of Sample Into PCR Reaction

2.1.1. Rationale for Loading Control

Quantitative RT-PCR without a control for loading has been described *(17)*. According to this method, quantified amounts of RNA are pipeted into each PCR reaction. However, there are two major quality control problems with this approach. First, there is no control for variation in RT from one sample to another and the effect will be the same as if unidentified, unquantified amounts of cDNA were loaded into the PCR reaction. It is possible to control for variation in RT by including a known number of internal standard RNA molecules in the RNA sample prior to RT *(18)*. However, as described in **Subheading 2.2.2.**, as long as there is control for the cDNA loading into the PCR, there is no need to control for variation in RT. Second, when gene expression values correlate to the amount of RNA loaded into the RT reaction, pipeting errors are not controlled for at two points. First, errors may occur when attempting to put the same amount of RNA from each sample into respective RT reactions. Second, if RT and PCR reactions are done separately, errors may occur when pipeting cDNA from the RT reaction into each individual PCR reaction. These sources of error may be controlled at the RNA level if an internal standard RNA for both a reference gene and each target gene were included with the sample prior to RT. However, this is a very cumbersome process and it limits analysis of the cDNA to the genes for which an internal standard was included. RT is most efficient and economical with at least 1 µg of total RNA. However, this amount of RNA would be sufficient for several hundred StaRT-PCR reactions and much of the RNA would be wasted if internal standards for only one or two genes were included prior to RT. Furthermore, internal standards must be within 10-fold ratio of the gene-specific native template cDNA molecules. It is not possible to know in advance the correct amount of internal standard for each gene to include in the RNA prior to RT so RT with a serial dilution of RNA would be necessary. Moreover, we, along with other investigators *(14)*, have determined that although RT efficiency varies from one sample to another, the representation of one gene to another in a sample does not vary among different reverse transcriptions and so internal standards are not necessary at the RNA extraction or RT steps. For these and other reasons, it is most practical to control for loading at the cDNA level.

2.1.2. Control for cDNA Loading Relative to Reference Gene

With real-time RT-PCR or StaRT-PCR, control for loading is best done at the cDNA level by amplifying a reference or "housekeeping" gene at the same time as the target gene. The reference gene serves as a valuable control for loading cDNA into the PCR reaction provided it does not vary significantly from the samples being evaluated.

2.1.3. Choice of Reference Gene

Many different genes are used as reference genes. No single gene is ideal for all studies. For example, β-actin varies little among different normal bronchial epithelial cell samples *(8)*, however it may vary over 100-fold in samples from different tissues, such as bronchial epithelial cells compared to lymphocytes. With StaRT-PCR it is possible to gain understanding regarding intersample variation in reference gene expression by measuring two reference genes, β-actin and glyceraldehyde-3-phosphate dehydrogenase (GAPDH), in every sample. We previously reported that there is a significant correlation between the ratio of β-actin/GAPDH expression and cell size *(5)*. This likely is a result of the role of β-actin in cytoskeleton structure. If the variation in reference gene

Table 1
Sources of Variation in Quantitative RT-PCR Gene Expression Measurement, and Control Methods

Source of Variation	Control Methods	
	StaRT-PCR[1]	Real-time
cDNA loading: Resulting from variation in pipeting, quantification, reverse transcription.	Multiplex Amplify with Reference Gene (e.g. β-actin)	Multiplex Amplify with Reference Gene (e.g. β-actin)
Amplification Efficiency		
Cycle-to-Cycle Variation: early slow, log-linear, and late slow plateau phases	Internal standard CT for each gene in a standardized mixture of internal standards (SMIS)	Real-time measurement
Gene-to-Gene Variation: in efficiency of primers	Internal standard CT for each gene in a SMIS	External standard curve for each gene measured
Sample-to-Sample Variation: variable presence of an inhibitor of PCR	Internal standard CT for each gene in an SMIS	Standard curve of reference sample compared to test sample[2]

Reaction-to-Reaction Variation: in quality and /or concentration of PCR reagents (e.g., primers)	Internal standard CT for each gene in a SMIS	None[2]
Reaction-to-Reaction Variation: in presence of an inhibitor of PCR	Internal standard CT for each gene in an SMIS	None[2]
Position-to-Position Variation: in thermocycler efficiency	Internal standard CT for each gene in an SMIS	None[2]

[1]StaRT-PCR involves (a) the measurement at end-point of each gene relative to its corresponding internal standard competitive template to obtain a numerical value, and (b) comparison of expression of each target gene relative to the β-actin reference gene, to obtain a numerical value in units of molecules/10^6 β-actin molecules. Use of references other than β-actin are discussed in text.

[2]With real-time RT-PCR, variation in the presence of an inhibitor in a sample may be controlled through use of standard curves for each gene in each sample measured and comparing these data to data obtained for each gene in a "calibrator" sample. However, variation in PCR reaction efficiency due to inhibitors in samples, variation in PCR reagents, or variation in position within thermocycler may be compensated only through use of an internal standard for each gene measured in the form of a SMIS. If an internal standard is included in a PCR reaction, quantification may be made at end-point, and there is no need for kinetic (or real-time) analysis. If internal standards for multiple genes are mixed together in a SMIS and then used to measure expression for both the target genes and reference gene, this is the patented StaRT-PCR technology, whether it is done by kinetic (real-time) analysis or at end-point. A SMIS fixes the relative concentration of each internal standard so that it cannot vary from one PCR reaction to another, whether in the same experiment, or in another experiment on another day, in another laboratory.

expression exceeds the tolerance level for a particular group of samples being studied, StaRT-PCR enables at least three alternative ways to normalize data among the samples, detailed in **Subheadings 2.1.4–2.1.6.**

2.1.4. Flexible Reference Gene

With StaRT-PCR, because the data are numerical and standardized owing to the use of a SMIS in each gene expression measurement, it is possible to use any of the genes measured as the reference for normalization. Thus, if there is a gene that appears to be less variable than β-actin, all of the data may be normalized to that gene by inverting the gene expression value of the new reference gene (to 10^6 β-actin molecules/molecules of reference gene) and multiplying this factor by all of the data, which initially are in the form of molecules/10^6 molecules of β-actin. As a result of this operation, the β-actin values will cancel out and the new reference gene will be in the denominator.

2.1.5. Interactive Gene Expression Indices

An ideal approach to intersample data normalization is to identify one or more genes that are positively associated with the phenotype being evaluated, and one or more genes that are negatively associated with the phenotype being evaluated. An interactive gene expression index (IGEI) is derived, comprising the positively associated gene(s) on the numerator and an equivalent number of the negatively associated gene(s) on the denominator. In these balanced ratios, the β-actin value is canceled. For example, this approach has been used successfully to identify an IGEI that accurately predicts anti-folate resistance among childhood leukemias *(11)*.

2.1.6. Normalization Against All Genes Measured

Because the data are standardized, if sufficient genes are measured in a sample, it is possible to normalize to all genes (similar to microarrays). The number of genes that must be measured for this approach to result in adequate normalization may vary depending on the samples being studied.

2.2. Control for Variation in Amplification Efficiency

PCR amplification efficiency may vary from cycle to cycle, from gene to gene, from sample to sample, and/or from well-to-well within an experiment.

2.2.1. Control for Cycle-to-Cycle Variation in Amplification Efficiency

PCR amplification rate is low in early cycles because the concentration of the templates is low. After an unpredictable number of cycles, the reaction enters a log-linear amplification phase. In late cycles, the rate of amplification slows as the concentration of PCR products becomes high enough to compete with primers for binding to templates. With StaRT-PCR *(5–15)*, as with other forms of competitive template RT-PCR *(3,17–20)* cycle-to-cycle variation in PCR reaction amplification efficiency is controlled through the inclusion of a known number of CT internal standard molecules for each gene measured. The ability to obtain quantitative PCR amplification at any phase in the PCR process, including the plateau phase, using CT internal standards has been confirmed by direct comparison to real-time RT-PCR *(22–24)*.

In contrast, with real-time RT-PCR, cycle-to-cycle variation in amplification efficiency is controlled by measuring the PCR product at each cycle, and taking the definitive measurement when the reaction is in log-linear amplification phase. A threshold fluorescence value known to be above the background and in the log-linear phase is arbitrarily established, and the cycle at which the PCR product crosses this threshold (C_T) is the unit of measurement *(25)*.

2.2.2. Control for Gene-to-Gene Variation in Amplification Efficiency

The efficiency of a pair of primers, as defined by lower detection threshold (LDT) cannot be predicted even after rigorous sequence analysis with software designed to identify those with the greatest efficiency. Based on extensive quality control experience developing gene expression reagents for more than 1000 genes, the LDT for primers thus chosen may vary more than 100,000-fold (from <10 molecules to 10^6 molecules). The only way to ensure that the LDT for a pair primers is below a desired level is to directly measure it with a known number of template molecules. The only way to do this for a human gene is to either PCR-amplify, synthesize, and/or clone a sufficient amount to quantify it. Once a sufficient amount has been prepared and quantified, it may be used in an external standard curve to determine LDT for real-time analysis, or as an internal standard to determine LDT by CT PCR. In StaRT-PCR an internal standard for each gene, in the form of a SMIS, is included in each gene expression measurement.

2.2.3. Control for Sample-to-Sample Variation in Amplification Efficiency

Variation in PCR amplification efficiency from sample-to-sample is often observed *(26)*, possibly resulting from variation in the presence of PCR reaction inhibitors, such as heme *(27,28)*. Importantly, amplification efficiency for different genes may be affected to different degrees in different samples *(26,29)*. In part for this reason, lacking proper controls comparison of the target gene to a reference gene will not be a reliable control for cDNA loading.

1. Internal Standards. With StaRT-PCR, the internal standard CTs control for variation in amplification efficiency, both among samples within a single experiment as well as among samples evaluated in multiple different experiments in different laboratories *(4–15)* (**Fig. 1**).

2. Standard Curve Comparison to Calibrator Samples. In contrast to StaRT-PCR, with real-time RT-PCR there is no internal control for intersample variation in PCR amplification efficiency. It is possible to achieve control by using a standard curve for the test sample and comparing these results to a standard curve for a calibrator sample *(29–31)*. However, standard curve measurements add time and expense to the real-time RT-PCR process. For each sample, it is necessary to do between 5 and 6 standard curve measurements along with measurement of the target gene. The standard curve should be run for each sample because intersample variation in amplification efficiency because of inhibitors is common and may alter the ΔC_T between a target gene and reference gene *(26)*.

3. Internal Standards in Real-Time. Theoretically, it would be possible to include internal standard CTs for both the target gene and reference gene in real-time PCR. For each gene, this would require preparation of one sequence-specific fluorescent probe for the NT and another for the CT. A probe specific to the NT would be homologous to the region that is in the NT but not in the CT. A probe specific to the CT would be homologous to the novel sequence formed when the reverse CT primer was incorporated (*see* **Subheading 3.2.2.** and **Fig. 2**). Real-time RT-PCR using an internal standard for a reference gene and a target gene in an SMIS would be StaRT-PCR, using a method other than densitometric measurement of electrophoretically separated bands to quantify the PCR products. If an SMIS were included in the PCR reaction, it no longer would be necessary to monitor the reaction in real-time, because quantification could be made relative to the internal standards at any point in the PCR amplification process, including end-point *(16,22–24,33)* (**Fig. 2**).

Fig. 2. (*Opposite page*) Simultaneous gene expression measurement by StaRT-PCR and real-time RT-PCR in two different samples. PCR amplification of a native template (NT) and respective internal standard competitive template (CT) for a target gene and reference gene (β-actin). Although StaRT-PCR NT and CT products routinely are quantified by densitometry at endpoint of PCR following electrophoretic separation (as represented by the bands labeled NT and CT) this schematic demonstrates how the reaction would look if measured at each cycle in real-time. For each real-time curve, the C_T is represented by a perpendicular black line. (**A**) For Sample 1, there were equivalent copies of β-actin NT and CT present at the beginning of the PCR reaction. Thus, following electrophoresis of the β-actin PCR products, the NT and CT bands are approximately equivalent and during real-time measurement, the fluorescent intensity for the NT will be about the same as for the CT. The NT/CT ratio is the same at an early cycle as it is at a late cycle (endpoint) even though the band intensity for both NT and CT is low at early cycle compared to late cycle. Similarly, the target gene NT band and CT band are about equivalent and the real-time value for the NT is about the same as for the CT. The ΔC_T between β-actin and the target gene is about 10. Methods for calcu-

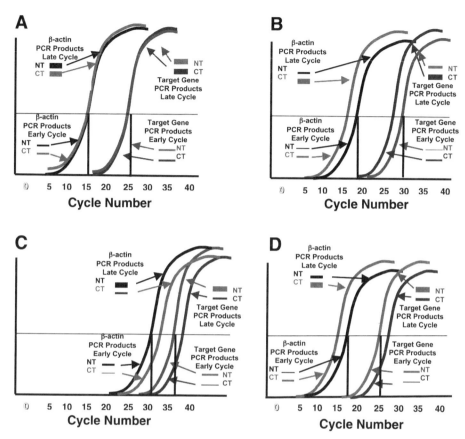

lating numeric value for target gene expression using StaRT-PCR are presented in Fig. 5 and **Subheading 3.8.** (**B**) For sample 2, the target gene is expressed at higher level than in sample 1. In addition, less cDNA was loaded into the PCR reaction and there were fewer NT then CT copies of β-actin present at the beginning of the PCR reaction. Thus, at the end of PCR the electrophoretically separated β-actin NT band is less dense than the CT band, and throughout real-time measurement the fluorescence value of the NT is less than that of the CT. However, even though less sample 2 cDNA was loaded into the PCR reaction, the target gene NT band is more dense than the target gene CT band, and the target gene NT fluorescence value during real-time measurement is higher throughout PCR and consequently, the ΔC_T is less than in sample 1, or about 7. (**C**) Repeat analysis of sample 1, but with low efficiency PCR. By real-time RT-PCR, ΔC_T is reduced from 10 to 6, characteristic of inhibitor in sample, inhibitor in well, or inappropriate concentration of reference gene primers and the result is artifactual. In contrast, by StaRT-PCR, there is no change in NT/CT ratio for either reference or target gene and result is the same as in absence of inhibitor. (**D**) Repeat analysis of sample 1, but with lower amount of cDNA loaded owing to variation in pipeting.

2.2.4. Control for Well-to-Well Variation in Amplification Efficiency

Possible sources of well-to-well variation in amplification efficiency include the presence of an inhibitor in some wells but not others, variation in the temperature cycling between different regions of a thermocycler block, or variation in concentration or quality of important reagents, such as primers. When one of these sources of variation markedly reduces PCR amplification efficiency in a well, it is possible that no PCR product will be observed in that well. Using real-time RT-PCR without internal standards in each PCR reaction, it is not possible to know whether to interpret absence or low level of PCR products as absence of transcript or inefficient PCR amplification (**Fig. 2**). An external standard curve would not be helpful because the PCR reactions would take place in different wells from the test sample. In contrast, using StaRT-PCR with internal standards in each PCR reaction, it is immediately possible to interpret the result correctly. The reagents for StaRT-PCR are carefully designed to amplify very efficiently so that for most genes a single molecule of CT or NT will be expected to give rise to detectable PCR product after taking stochastic issues into consideration. The lowest concentration of CT molecules present in a StaRT-PCR reaction is 10^{-17} M with Mix F (*see* **Subheading 3.4.**).

In a 10 µL PCR reaction volume 10^{-17} M represents 60 molecules. With 60 molecules of internal standard present in the PCR reaction and all of the components of the PCR reaction functioning properly, if a gene is not expressed in a sample, the PCR product for the internal standard will be observed but the PCR product for the NT will not. One can then conclude that the gene expression was so low that for cDNA included in the PCR reaction there was less than six molecules (10-fold less than the number of CT molecules) of cDNA representing that gene. On the other hand, if neither NT nor CT product is detectable, the PCR reaction efficiency was suboptimal and no interpretation can be made regarding level of expression.

2.3. Schematic Comparison of StaRT-PCR to Real-Time RT-PCR

In **Fig. 2** is a schematic presentation of the way quantitative measurements are made in the two forms of quantitative RT-PCR discussed here; real-time RT-PCR and StaRT-PCR. In real-time, the fluorescent PCR product is measured at each of 35–40 cycles. As many as four PCR products may be monitored simultaneously in real-time if four different fluors are used. In **Fig. 2A**, the NT and CT for β-actin and the NT and CT for the target gene are PCR-amplified simultaneously.

In StaRT-PCR, the products of endpoint PCR are electrophoretically separated and the shorter CT PCR product migrates faster than the NT PCR product. The PCR products are electrophoresed in the presence of fluorescent intercalating dye and densitometrically quantified. If there is more NT product than CT product, the NT band will emit a more intense fluorescent light. If there is more CT product than NT product, the CT band will be brighter. Importantly, the ratio of NT/CT that is present at the beginning of PCR will remain constant throughout PCR to endpoint. For this reason, with StaRT-PCR it is not necessary to monitor the PCR reaction in real-time to ensure that the reaction is in log-linear phase (**Fig. 2A**). In addition, measurement of both a reference and a target gene in every PCR reaction controls for loading from one sample to another (**Fig. 2B**) or among replicate measurements of the same sample (**Fig. 2D**). With StaRT-PCR, variation in PCR amplification efficiency caused by the presence of an inhibitor in the sample, an inhibitor in the PCR reaction vessel, defective PCR reagent, or wrong concentration of a PCR reagent is controlled for by the presence of internal standards in every PCR reaction.

With real-time RT-PCR, it is possible to control for loading by measuring the target gene and reference gene in the same PCR reaction (**Fig. 2A,B,D**). The C_T for the reference gene and the target gene both may vary from one experiment to another, but the ΔC_T will not vary. However, real-time may not control for well-to-well variation in the quality or quantity of PCR reagents, or sample-to-sample variation in PCR efficiency resulting from the presence of inhibitors, for example, heme **Fig. 2C**). Presence of an inhibitor may lead to variation in PCR amplification efficiency of one gene compared to another *(26)*. A bad lot or inappropriate concentration of primers for the reference gene or the target gene would cause variation in PCR amplification of one gene relative to another. As depicted here, (**Fig. 2C**), amplification efficiency of the reference gene in sample 1 is affected by low concentration of primer, but amplification efficiency of the target gene is normal. The result is that the ΔC_T is reduced from ten in **Fig. 2A** to six in **Fig. 2C**, and the value for expression of the target gene is inappropriately high. In contrast, for StaRT-PCR because the amplification efficiency of the internal standard is affected the same way as the NT for each gene, the ratio is unchanged in **Fig. 2A,C** for either reference gene or target gene, and using the ratio of NT/CT for target gene relative to NT/CT for reference gene controls for variation in amplification efficiency. See **Subheadings 3.6–3.8.** for details of how StaRT-PCR data are calculated.

3. StaRT-PCR Method

3.1. Materials

1. StaRT-PCR reagents, including primers and SMIS are purchased from Gene Express, Inc. (GEI, Toledo, OH).
2. Buffer for Idaho Rapidcycler air thermocycler: 500 mM Tris-HCl, pH 8.3, 2.5 µg/µL BSA, 30 mM MgCl$_2$ (Idaho Technology, Inc., Idaho Falls, ID).

3. Buffer for block thermocyclers, Thermo 10 X, 500 mM KCl, 100 mM Tris-HCl, pH 9.0, 1.0% Triton X-100 (Promega, Madison, WI).
4. *Taq* polymerase (5U/µL), Moloney Murine Leukemia Virus (MMLV) reverse transcriptase, MMLV RT 5X first strand buffer: 250 mM Tris-HCl, pH 8.3, 375 mM KCl, 15 mM MgCl$_2$, 50 mM dithiothreitol, oligo dT primers, Rnasin, pGEM size marker, and deoxynucleotide triphosphates (dNTPs) also are obtained from Promega.
5. TriReagent is obtained from Molecular Research Center, Inc. (Cincinnati, OH).
6. Ribonuclease (Rnase)-free water and TOPO TA cloning kits are obtained from Invitrogen (Carlsbad, CA) (*see* **Note 1**).
7. GigaPrep plasmid preparation kits are purchased from Qiagen (Texas).
8. Caliper AMS 90SE chips are obtained from Caliper Technologies, Inc. (Mountain View, CA).
9. DNA purification columns were obtained from QiaQuick (Qiagen, Valencia, CA).

3.2. Methods

3.2.1. RNA Extraction and Reverse Transcription

1. RNA Extraction and Quantification: Pellet the cell suspensions, pour off the supernatant, and dissolve the pellet in TriReagent and extract according to manufacturer's instructions and previously recorded methods (**32**). Store the RNA pellet under ethanol at −80°C, or suspend in RNASe free water, and freeze at −80°C. It may be safely stored in this condition for years. Evaluate the quality of the RNA on an Agilent 2100 using the RNA chip, according to manufacturer's instructions.
2. Reverse Transcription: Reverse transcribe 1 µg total RNA using MMLV RT and an oligo dT primer as previously reported (**35**). For small amounts of RNA (e.g. < 100 ng), the efficiency of reverse transcription is better with Sensicript™ than with MMLV reverse transcriptase. We have obtained efficient RT from as little as 50 ng of RNA with Sensicript™. Incubate the reaction at 37°C for 1 h.

3.3. Synthesis and Cloning of Competitive Templates (see Note 2)

3.3.1. Native Template Primer Design

Before constructing the CT for each gene, the primer pair must efficiently amplify the native cDNA. Design primers with the following characteristics:

1. Amplify from 200 to 850 bases of the coding region of targeted genes
2. Annealing temperature of 58°C (tolerance of +/−1°C) (*see* **Note 3**).

3.3.2. Native Template Primer Testing

Design primers according to above steps, synthesize and use to amplify native template in appropriate cDNA sample. The presence of a single strong band after 35 cycles of PCR is verification that the primers are efficient and specific (*see* **Note 4**).

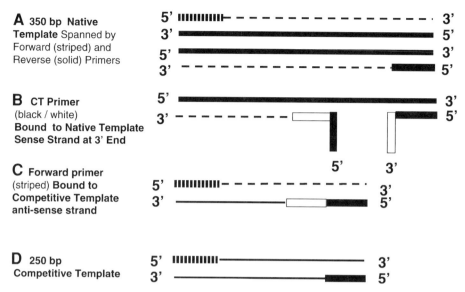

Fig. 3. Preparation of internal standard competitive templates. (**A**) Forward (striped bar) and reverse (black bar) primers (approx 20 bp in length) that span a 150–850 bp region are used to amplify the native template (NT) from cDNA. *Taq* polymerase will synthesize NT DNA from these primers (dashed lines). (**B**) After confirming that native template primers work, a CT primer is designed. This is an approx 40 bp primer with the sequence for the reverse primer (black bar) at the 5' end, and a 20 bp sequence homologous to an internal native template sequence (white bar) at the 3' end, collinear with the reverse primer sequence. The 3' end of this 40 bp primer is designed to be homologous to a region approx 50–100 bp internal to the reverse primer. The 5' end of this 40 bp primer will hybridize to the region homologous to the reverse primer, while the 3' end will hybridize to the internal sequence. Importantly, *Taq* polymerase will be able to synthesize DNA using only the primers bound at the 3' end (dashed line). (**C**) In the next cycle of PCR, the DNA newly synthesized using the 40 bp primer hybridized to the internal sequence is bound to forward primer (striped bar), and a homologous strand is synthesized. (**D**) This generates a double stranded CT with the reverse primer sequence 100 bp closer to the forward primer than occurs naturally in the NT. This method is as previously described *(34)*.

3.4. Competitive Template Primer Design

After suitable primers for NT amplification have been designed and tested, prepare a CT primer according to previously described methods *(36)*, as schematically presented in **Fig. 3**.

1. Competitive Template Primer Testing. The 40 bp CT primer is paired with the forward primer designed in **Subheading 3.3.1.** and used to amplify CT from native cDNA.

3.5. Competitive Template-Internal Standard Production

1. For each gene, set up five 10 μL PCR reactions using the native forward primer and the CT primer and amplify for 35 cycles.
2. Combine the products of these five PCR reactions, electrophorese on a 3% NuSieve gel in 1X TAE, and cut the band of correct size from the gel and extract using the QiaQuick method.
3. Clone the purified PCR products into PCR 2.1 vector using TOPO TA cloning kits then transform into HS996 (a T1-phage resistant variant of DH10B).
4. After cloning, transformation, and plating on LB plates containing X-Gal, IPTG, and carbenicillin, pick three isolated white colonies. Prepare plasmid minipreps, perform*EcoRI* digestion and electrophorese on 3% SeaKem agarose. For those clones documented to have an insert by *EcoRI* digestion, confirm the insert to be the desired one by sequencing the same undigested plasmid preparation using vector specific primers. Only those clones with homology to the correct gene sequence and that have 100% match for the primer sequences proceed to large-scale CT preparation and are included in the standard mixes. Those that pass this quality control assessment then continued to the next steps.
5. Prepare each quality assured clone in quantities large enough (1.5 L) to allow for <1 billion assays (approx 2.6mg).
6. Purify plasmids from resultant harvested cells using Qiagen GigaPrep kits.
7. Carefully quantify plasmid yields using a Hoeffer DyNAQuant 210 fluorometer.
8. For each CT that passes all of the defined quality control steps described in **step 4**, assess the sensitivity of the cloned CT and primers by performing PCR reactions on serial dilutions and determine the limiting concentration that still yield a PCR product. Only those preparations and primers that allow for detection of 60 molecules or fewer (a product obtained with $10^{-17} M$ CT in 10 μl PCR reaction volume) are continued for inclusion into SMIS (*see* **Note 5**).

3.6. Preparation of Standardized Mixtures of Internal Standards (SMIS) (see Note 6)

Combine cloned and quantified CTs into SMIS according to modifications of previously described methods *(5,6,36)*.

1. Mix plasmids from quality assured preparations (see **Subheading 3.4.**) into SMIS representing 24 genes.
2. The concentration of the competitive templates in the 24 gene SMIS is $4 \times 10^{-9} M$ for β-actin CT, $4 \times 10^{-10} M$ for GAPD (CT1), $4 \times 10^{-11} M$ for GAPD (CT2), and $4 \times 10^{-8} M$ for each of the other CTs (*see* **Note 7**).
3. Linearize each 24 gene SMIS by NotI digestion. Incubate the SMIS with *NotI* enzyme at a concentration of 1 unit/μg of plasmid DNA in approx 15 mL of buffer at 37°C for 12–16 h.
4. Combine four linearized 24-gene SMIS in equal amounts to yield 96-gene CT mixes with a maximum concentration of $10^{-9} M$ for β-actin, $10^{-10} M$ GAPD (CT1), $10^{-11} M$ GAPD (CT2), and $10^{-8} M$ for the other CTs.

5. Serially dilute high concentration SMIS with a reference gene CT mixture comprising β-actin CT (10^{-9} *M*) and two different GAPD CTs, GAPD CT1 (10^{-10} *M*), and GAPD CT2 (10^{-11} *M*). This yields six stock SMIS (A–F) with β-actin, GAPD1 and GAPD2 at constant concentrations of 10^{-9} *M*, 10^{-10} *M*, and 10^{-11} *M* respectively while the concentration of the other CTs in SMIS A–F respectively are 10^{-8} *M*, 10^{-9} *M* 10^{-10} *M*, 10^{-11} *M*, 10^{-12} *M*, and 10^{-13} *M*.

6. Dilute stock concentration SMIS 1000-fold to working solutions containing β-actin, GAPD1 and GAPD2 at concentrations of 10^{-12} *M*, 10^{-13} *M*, and 10^{-14} *M* respectively while the concentration of the other CTs in SMIS A–F respectively are 10^{-11} *M*, 10^{-12} *M* 10^{-13} *M*, 10^{14} *M*, 10^{-15} *M*, and 10^{-16} *M*.

3.7. StaRT-PCR

StaRT-PCR is performed using previously published protocols *(5,6)*. StaRT-PCR is performed using previously published protocols *(5,6)*. First, the cDNA sample is diluted until 1 μL competes equally with 6×10^5 molecules of β-actin CT (1 μL of SMIS containing 10^{-12} *M* β-actin CT). The NT/CT must be greater than 1:10 and less than 10:1 for the measurement to be within linear dynamic range. Typically, this is the amount of cDNA derived from 100 to1000 cells. Next, this amount of cDNA sample is PCR amplified in multiplex with a SMIS containing internal standards for reference genes and target genes and gene specific primers from Gene Express, Inc. as described earlier. As with the reference gene, the target gene NT/CT must be greater than 1:10 and less than 10:1. Because genes are expressed over more than six orders of magnitude, this explains why the target gene CTs in each 96-gene SMIS must be 10-fold serially diluted relative to the reference gene CTs, in mixes A–F. For each 96-gene SMIS, sufficient amount of A–F mix is prepared for more than 100 billion assays. Thus, these SMIS are constant and may be used by all labs. This is schematically represented in **Fig. 4**. In **Fig. 4,** genes 6 and 7 are expressed at a low level in sample A and therefore are measured using SMIS E. In sample B, genes 6 and 7 are expressed at a higher level and are measured using SMIS C and D, respectively. All of the values can be compared because all of the SMIS are standardized and constant. For each experiment, a PCR master mixture is prepared containing the appropriate amount of cDNA and SMIS for the number of gene expression assays to be done. Next, the reference gene NT is measured relative to its CT, and the target gene is measured relative to its CT, and expression is calculated as target gene molecules/10^6 β-actin molecules. Briefly, StaRT-PCR is done by a) including in each PCR reaction a sample of cDNA and a known amount of SMIS, and b) multiplex RT-PCR amplifying both the target gene NT and its respective CT and a reference gene (e.g., β-actin) NT and its respective CT for every gene expression measurement (**Figs. 1,3**). These four templates may be amplified in the same tube *(4,5)* or, if the experiment is

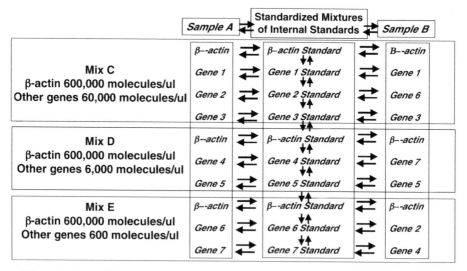

Fig. 4. Relationship among mixes serially 10-fold diluted from each 96-gene SMIS. As described in text, a serial 10-fold dilution, A–F, of target gene internal standards relative to reference gene internal standards is prepared for each 96-gene SMIS. This allows StaRT-PCR measurement of each gene, even though different genes may be expressed over a range of more than 6 orders of magnitude.

properly designed, the NT and CT pair for the target gene and the NT and CT pair for the reference gene may be amplified in separate tubes *(5)*.

3.8. Step-by-Step Description of StaRT-PCR Method

1. Balance cDNA with 6×10^5 β-actin CT molecules (the amount of β-actin CT in 1 μL of SMIS). After establishing the amount of cDNA in balance with 6×10^5 copies of β-actin CT, this amount of cDNA is used in all subsequent experiments (*see* **Note 8**).

2. Combine and mix a volume of cDNA sample (diluted to the level that is in balance with the amount of β-actin CT in 1 μL of SMIS (6×10^5) molecules, as determined above) with an equal volume of the appropriate SMIS A–F such that the target gene NT/CT will be greater than 1/10 and less than 10/1. A 1 μL volume of each is used for each gene expression assay to be performed (*see* **Note 9**). If the appropriate SMIS is not known for a particular gene in a sample from a particular type of tissue, expression is measured in both SMIS C and E. This allows measurement over four orders of magnitude. For the few genes expressed at very high or low level, it will be necessary to repeat analysis with SMIS A or F. In the SEM Center, described later, the most appropriate SMIS is selected based on data in the standardized expression database.

3. Combine cDNA/SMIS mixture from previous step with other components of the PCR reaction mixture (buffer, dNTPs, Mg++, *Taq* polymerase, H_2O)
4. Prepare tubes or wells with a primer pair for a single gene. If products are to be analyzed by PE 310 device (*see* **Subheading 3.4.9.**) the primers should be labeled with appropriate fluor.
5. Place aliquots of this PCR reaction mixture into individual tubes each containing primers for a single gene (*see* **Note 10**).
6. PCR Amplification. Cycle each reaction mixture either in an air thermocycler (e.g., Rapidcycler (Idaho Technology, Inc., Idaho Falls, ID) or block thermocycler (e.g., PTC-100 block thermal cycler with heated lid, MJ Research, Inc., Incline Village, NV; laboratories) for 35 cycles. In either thermocycler, the denaturation temperature is 94°C, the annealing temperature is 58°C, and the elongation temperature is 72°C.
7. Separation and Quantification of NT and CT PCR Products (*see* **Note 11**).
    a. Agarose gel. Following amplification, load the entire volume of PCR product (typically 10 µL) into wells of 4% agarose gels (3/1 NuSieve: SeaKem) containing 0.5 µg/mL ethidium bromide. Electrophorese gels for approx 1 h at 225 V in continuously chilled buffer, then visualize and quantify with an image analyzer (products available from Fotodyne, BioRad).
    b. PE Prism 310 Genetic Analyzer CE Device. Amplify PCR products with fluor-labeled primers. One microliter of each PCR reaction is combined with 9 µL of formamide and 0.5-0.1 µL of ROX size marker. Heat samples to 94°C for 5 min and flash cooled in an ice slurry. Load samples onto the machine and electrophorese at 15 kV, 60°C for 35–45 min using POP4 polymer and filter set D. The injection parameters are 15 kV, 5 sec. Fragment analysis software, GeneScan (Applied Biosystems, Inc., Foster City, CA) is used to quantify peak heights that are used to calculate NT/CT ratios. No size correction is performed since each DNA molecule was tagged with one fluorescent marker from one labeled primer.
    c. Agilent 2100 Bioanalyzer Microfluidic CE Device. The DNA 7500 or DNA 1000 LabChip kit may be used. Following amplification, load 1 µL of each 10 µL PCR reaction into a well of a chip prepared according to protocol supplied by manufacturer. Run DNA assay, which applies a current to each sample sequentially to separate NT from CT. DNA is detected by fluorescence of an intercalating dye in the gel-dye matrix. NT/CT ratios are calculated from area under curve (AUC) and a size correction is made.
    d. Caliper AMS 90 Microfluidic CE Device. Set up the PCR reactions in wells of a 96- or 384-well microplate. Following amplification, place the microplate in the Caliper AMS 90. Follow the protocol recommended by the manufacturer. The AMS 90 removes and electrophoreses a sample from each well sequentially every 30 sec. The NT and CT PCR products are separated and quantified. Because detection is through fluorescent intercalating dye, size correction is necessary.
    e. MALDI-TOF separation. A method for separating PCR products recently was described (*16*). This method may be applied to analysis of StaRT-PCR products resulting from amplification of cDNA in the presence of SMIS.

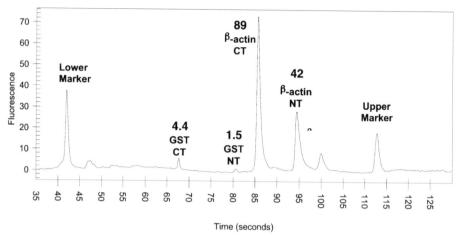

Fig. 5. Calculations involved in StaRT-PCR measurement of GST gene expression relative to β-actin in an actual bronchial epithelial cell (BEC) sample. The native template (NT) PCR product was amplified from cDNA specific for the gene being measured, and the competitive template (CT) PCR product was amplified from the internal standard for each respective gene. A volume of SMIS containing a known number of internal standard CT molecules for β-actin (600,000) and GST (6000) were included at the beginning of the PCR reaction. For each gene the NT and CT will amplify with the same efficiency. Thus, the β-actin gene NT/CT PCR product ratio allows determination of the number of β-actin NT copies at the beginning of PCR and the target gene NT/CT ratio allows determination of the number of target gene NT copies at the beginning of PCR. See text for steps used to calculate gene expression values.

3.9. Steps to Calculate the Number of NT Molecules Present at the Beginning of PCR for Each Gene

Calculation of gene expression. Values are calculated in units of target gene cDNA molecules/10^6 β-actin cDNA molecules. The steps taken to calculate gene expression are based on densitometric measurement values for the electrophoretically separated NT and CT PCR products such as those presented in **Fig. 5**. The calculations below are based on the example in **Fig. 5**.

1. Correct NT PCR product area under the peak (AUP) to length of CT DNA.
2. Determine ratio of corrected NT AUP relative to CT AUP.
3. Multiply NT/CT value × number of CT molecules at beginning of PCR.
4. Calculation of reference gene (β-actin) molecules using above protocol.
    a. 416/532(β-actin CT bp/ NT bp) × 42 (NT AUP) = 33 (corrected NT value).
    b. Correct β-actin NT AUP divided by β-actin CT AUP = 0.37.
    c. 0.37 (β-actin NT/CT) × 600,000 (number of β-actin CT molecules at beginning of PCR) = 222,000 NT molecules at beginning of PCR.

5. Calculation of target gene (GST) molecules using above protocol:
   a. 227/359 (GST CT bp/NT bp) × 1.5 (NT AUP) = 0.95 (corrected NT AUP).
   b. 0.95 (GST corrected NT AUP) divided by 4.4 (GST CT AUP) = 0.22.
   c. 0.22 (GST NT/CT) × 6000 (number of GST CT molecules at beginning of PCR) = 1290 GST NT molecules at beginning of PCR.
6. Calculation of molecules of GST/10^6β-actin molecules
   1290 GST NT molecules/222,000 β-actin NT molecules = 580 GST molecules/106 β-actin molecules.

4. The Standardized Expression Measurement Center

The SEM Center was recently established at the Medical College of Ohio through a grant from the National Cancer Institute. The SEM Center is in operation and available for use at www.geneexpressinc.com.

Currently, microarray technology is the starting point for most large-scale gene expression profiling investigations. However, owing to limits in lower detection threshold and sensitivity, and lack of internal standards, microarray technology is most appropriately applied as a screening tool. For most applications, data obtained through microarray analysis must be validated by a more sensitive and quantitative method. Most investigators use a quantitative RT-PCR method for this purpose.

The purpose of the SEM Center is to provide standardized, reproducible, gene expression measurement. The SEM Center achieves these goals by using StaRT-PCR. Further, StaRT-PCR is easily automated and subjected to quality control, which is critical for analysis of clinical specimens.

The SEM Center function is similar to that of a DNA sequencing service. Thus, users send their RNA or cDNA samples to the SEM Center for analysis. Users select a set of genes for measurement and send a requisition listing these selected genes (available at the SEM Center website) along with the samples.

4.1. Technology Incorporated by the SEM Center

A PE Robotic liquid handler is used to prepare 10 μL PCR reactions in 96-well or 384-well microplates. First, the liquid handler is programmed to distribute 1 μL of primers for the requested genes into wells of the microplates. Second, for each cDNA a sufficient volume of PCR mixture for the anticipated number of gene expression measurements is prepared, containing buffer, *Taq* polymerase, dNTPs, cDNA and internal standards. The robot then distributes 9 μL of this PCR reaction mixture into each well. Thus, in each well the internal standard CTs for each gene and cDNA are present in the same ratio, however, because only one pair of primers is present in each well, only one gene and its respective internal standard CT are amplified in each well. Following 35 cycles of PCR, each microplate is transferred to the Caliper AMS 90 for analysis.

When StaRT-PCR was first developed, products were separated on agarose gels *(4,5)*. This method is reliable but relatively costly, time consuming, and labor intensive. Through advances in capillary electrophoresis (CE), alternative methods for separation of StaRT-PCR products that are faster and less expensive have become available. We compared separation of StaRT-PCR products on agarose gel, PE 310 CE, and Agilent 2100 Bioanalyzer mcrofluidic CE *(31)*. Each of these methods provided the same, reproducible results. Theoretically, the internal standard mixtures prepared for StaRT-PCR may be used to measure gene expression coupled with any method capable of quantifying strands of DNA with different sizes, including HPLC and mass spectrometry. Quantification of gene expression through analysis of RT-PCR products by MALDI-TOF MS has been recently described *(16)*.

Currently, the Caliper AMS 90 is used for high-throughput separation of StaRT-PCR products in the SEM Center. This device is capable of 1000 gene expression assays in eight hours. The SEM Center employs a microfluidic chip with a sipper that moves from well to well of a microplate, aspirating and then electrophoretically separating StaRT-PCR products every 30 s. This allows analysis of a 384-well plate in approx 3 h, which is comparable to the throughput of the fastest real-time devices.

4.2. Design of High-Throughput StaRT-PCR Experiments

All of the genes that are to be measured in a given sample are measured simultaneously. Owing to the presence of SMIS in every PCR reaction, gene expression values for one sample may be compared to gene expression values from another sample and evaluated at a different time **(Fig. 1A)**.

PCR products (NT and CT) for as many as four genes may be electrophoresed (separated and quantified) in the same microfluidic channel of the AMS 90SE. Accomplishing this in the high-throughput SEM Center requires software that identifies genes that may be electrophoresed simultaneously, based on the length in base pairs (bp) of the NT and CT PCR products. As described in **Subheading 3.** for each gene, the primers and CTs are designed to amplify PCR products that range from 150–850 bp. Thus, for every set of genes to be analyzed, the software must identify which genes may be electrophoresed together.

4.3. Use of Multiplex StaRT-PCR to Reduce cDNA Consumption

An advantage of quantitative RT-PCR as a tool for measuring gene expression is that it consumes very small amounts of cDNA. This enables meaningful analysis of very small-tissue biopsy samples, such as those obtained by fine-needle aspirate. Despite the low amount of cDNA required in quantitative RT-PCR, high-throughput analysis of many genes simultaneously will consume large amounts of cDNA for each sample, possibly limiting the analysis of small samples. However, multiplex StaRT-PCR methods recently described *(7)* may solve this problem. It should be possible to combine nanotechnology methods for manipulating small liquid volumes with multiplex StaRT-PCR methods to decrease the PCR reaction volumes to 10–100 nL.

The multiplex StaRT-PCR method involves two rounds of PCR. In the first round, cDNA, CT mix, and primers for up to 96 genes are amplified for 35 cycles. Next, PCR products from round one are diluted, combined with primers for one gene, and amplified for an additional 35 cycles. No additional CT or cDNA is added. Products from round one may be diluted as much as 100,000-fold and still be quantified following round two amplification. Thus, using multiplex StaRT-PCR, 96 genes may be measured in the same amount of cDNA typically used to measure one gene with uniplex StaRT-PCR. The gene expression values obtained for multiplex StaRT-PCR are highly correlated with those obtained by uniplex StaRT-PCR *(7)*.

Multiplex StaRT-PCR works because gene expression measurements are determined by the ratio of NT/CT for each gene and not by the absolute amount of NT PCR product. For each gene, NT and CT are amplified with the same primers, share sequence homology, and amplify with equal efficiencies *(7)*. Therefore, differences in amplification efficiency will not affect the measured relative level of expression between genes in different samples even after two rounds of amplification.

4.4. Other SEM Center Services

The SEM Center provides other services besides gene expression measurement, and these are listed on the requisition that may be downloaded from www.geneexpressing.com. Users may submit cDNA or RNA samples. RNA samples will be assessed for quality on an Agilent 2100 RNA chip. If the RNA quality is good, it will be reverse transcribed. The amount of cDNA produced will be quantified by measuring the number of β-actin molecules in a serially diluted sample. If sufficient cDNA is present for the requested number of gene expression measurements, the SEM Center will proceed with the order. If there is insufficient amount of cDNA, the user will be notified and asked to prioritize genes to be measured, or send more RNA or cDNA.

4.5. Standardized Gene Expression Database

Users send samples to the SEM Center without any annotating information and with a requisition that includes an attestation that any primary human samples were obtained under approved and active IRB protocol. Because no potentially identifying information is provided, the SEM Center is exempted from the need to obtain an Institutional Review Board protocol for each set of samples submitted. As soon as an order is completed, the data are sent by email and a hard copy sent to the user. Each user is encouraged to send the annotating information as soon as possible. It is hoped that users will send the annotating information as soon as a manuscript containing the data is accepted for publication, or sooner. An annotated standardized gene expression database will be key for advances in research as well as for developing clinical tests.

5. Notes

1. The quality of the RNase-free water is critical to efficient extraction of intact RNA. We have found that it is more cost effective to purchase reliable RNase-free water from commercial sources than it is to prepare our own. Either inadequate DEPC treatment or inadequate removal of DEPC after treatment can inhibit reverse transcription and PCR (*see* **Subheading 3.1.6.**).
2. Internal standard CTs are constructed by Gene Express, Inc. (GEI, Toledo, OH) based on previously described methods (*5,6,36*) (*see* **Subheading 3.3.**).
3. Use Primer 3.1 software (Steve Rozen, Helen J. Skaletsky, 1996, 1997) Primer 3. Code available at http://www-genome.wi.mit.edu/genome_software /other/primer3. html) to design primers. Designing primers with the same annealing temperature allows StaRT-PCR reactions to achieve approximately the same amplification efficiency under identical conditions. If there is variation in amplification efficiency it does not cause variation in quantitative value because the value is obtained from the ratio between the NT and CT for the same gene, and amplification efficiency of the NT and CT for the same gene are affected identically.

   Designing primers that amplify different sized products for different genes will support automation and high-throughput applications, including capillary gel and microchannel CE. Primer sequences and Genbank accession numbers for genes designed by GEI are available at www.geneexpressinc.com. (*see* **Subheading 3.3.1.**).
4. Primers are tested using reverse transcribed RNA from a variety of tissues or individual cDNA clones known to represent the gene of interest. Primer pairs that fail to amplify the target gene in any tissue or individual cDNA clone (less than 10% of the time) are redesigned and the process repeated (*see* **Subheading 3.3.2.**).
5. The number of molecules at different molarities is a multiple of six as a consequence of Avogadro's Number (6.02×10^{23} molecules/mole). More than 80% of the CTs developed have a sensitivity of six molecules or less. Thus, for these genes, it is possible to measure as few as 10 molecules/ 10^6 β-actin molecules. Because there are approximately 100–1000 β-actin molecules per cell for most cell types, this level of sensitivity allows measurement of 1 molecule per 100–1000 cells. At the other end of the expression spectrum, SMIS A will allow measurement of more than 10^7 molecules/10^6 molecules of β-actin (10^3–10^4 molecules/cell). In our experience, few genes approach this level of expression, examples include UGB (Genbank no. U01101) and vimentin (X56134) (unpublished data). Thus, SMIS A–F should allow measurement of gene expression over the full spectrum observed in human tissues (*see* **Subheading 3.6.**).

6. The process of identifying primers that lead to high PCR amplification efficiency for both the NT and CT, preparing large amounts of the CT through cloning, quantifying the CTs, and mixing the CTs into SMIS, transforms CTs into internal standards. Thus, CTs are the raw material necessary for development of the much more valuable product (*see* **Subheading 3.6.**).

7. The reason for two different GAPD CTs is that the expression of GAPD relative to β-actin may vary as much as 100-fold from one tissue type to another. Having two different concentrations of GAPD CT relative to β-actin enables comparison of GAPD to β-actin in all samples. These comparisons are helpful in determining intersample variation in expression of reference genes (*see* **Subheading 3.7.**).

8. For each cDNA sample, it is necessary to determine the dilution of the test cDNA that is approximately (within 10-fold range) in balance with 600,000 copies of β-actin (1 μL of SMIS containing β-actin CT at 10^{-12} *M*). This is approximately the amount of cDNA derived from 100 to 1000 cells. This amount will ensure that there is sufficient cDNA to quantify genes expressed at low levels. If the goal is to have at least 10 transcripts present at the beginning of PCR to avoid stoichiometric problems, this amount of cDNA will allow quantification of genes expressed as low as 1 transcript in every 10–100 cells. If less sensitivity is required, less cDNA may be used. Thus, one could choose to use the amount of cDNA in balance with 60,000 molecules of β-actin CT. This will not allow measurement of genes expressed at very low levels, but will be sufficient for analysis of most genes and will reduce consumption of cDNA 10-fold. This may be useful when analyzing very small biopsy specimens for diagnostic tests. For each of the SMIS A–F, 1 μl of CT mix contains 600,000 molecules of β-actin CT, thus any of the SMIS could be used for this purpose of balancing cDNA with β-actin. The standard operating procedure is to use SMIS F.

   A common mistake for beginning users of StaRT-PCR is to balance the cDNA with the β-actin in the SMIS initially, and then, when the target gene NT and CT are not in balance, vary the amount of cDNA in the PCR reaction mixture to get the target gene NT/CT in balance. Instead, keep the amount of cDNA constant and change the SMIS used. The SMIS have been prepared for measurement of genes across the full range of gene expression measurement (6 orders of magnitude). Because the NT/CT ratio must be within 10-fold ratio in order to obtain reliable, reproducible quantification, six different SMIS have been prepared, containing 10-fold serial dilution of all target gene CTs relative to reference gene CT. If SMIS D were used to measure a target gene, and the target gene NT was more than 10-fold greater than the CT, the next step would be to repeat the experiment with the same amount of cDNA, but using SMIS C, which has a 10-fold higher concentration of target gene CT (*see* **Subheading 3.8.**).

9. The StaRT-PCR method standardizes every gene expression measurement so that it can be readily compared to all other StaRT-PCR measurements. The procedure described in this step allows one to compare the NT/CT ratio for the reference gene to the NT/CT ratio for the target gene in a reliable way that controls for variation in pipeting. This step commonly is carried out incorrectly by users of StaRT-PCR.

For example, it is common for users to aliquot SMIS sufficient for a single gene expression measurement into each separate PCR reaction mixture, and then aliquot cDNA for a single measurement into each tube. Owing to pipeting errors, this would be associated with variation in the NT/CT ratio of each target gene relative to the NT/CT ratio for the reference gene, as well as that for other target genes.

The SMIS (A, B, C, D, E, or F) selected will be the one containing CT at the concentration most likely, based on previous experience, to be in balance (within 10-fold range) with the gene or genes being assessed (*see* **Subheading 3.8.2.**).

10. In **Subheading 3.8.5.** of this experimental design, the ratio of CT for every gene in the mixture relative to its corresponding NT in the cDNA is fixed simultaneously. When aliquots of this mixture are transferred to PCR reaction vessels, although variations in loading volumes resulting from pipeting errors are unavoidable, there is no potential for variation in any target gene NT/CT ratio relative to reference gene NT/CT ratio. In addition, it enables standardized expression measurement. In order to ensure control for loading in each experiment, the reference gene (β-actin) is measured along with the target genes for each different master mix utilized. The choice of which four SMIS to use is based on previous experience. For example, if among all previous samples a gene has been expressed within a range of 10^1–10^3 molecules/10^6 β-actin molecules, the gene will be measured using SMIS E. In contrast, if among all previous samples, a gene has been expressed within a range of 10^5–10^7 molecules/10^6 β-actin molecules, the gene will be measured using SMIS B. For the rare samples that express the gene outside of the expected ranges, a follow-up analysis with the appropriate CT mix is performed.

11. Electrophoresis may occur in an agarose gel, capillary electrophoresis device (e.g., PE 310), or microfluidic CE device (e.g., Agilent 2100 or Calipertech AMS 90 high-throughput system). If an agarose gel is used, electrophoresis is for one hour at 225 V through agarose gel. If a CE device or microfluidic CE device is used, electrophoresis is according to the manufacturer's instructions. Following electrophoresis, the relative amount of NT and CT is determined by densitometric quantification of bands that have been stained by an intercalating dye (e.g., ethidium bromide). Theoretically, the internal standard mixtures prepared for StaRT-PCR may be used to measure gene expression using any method capable of quantifying strands of DNA with different sizes and/or sequence, including solid phase hybridization MALDI-TOF and HPLC (*see* **Subheading 3.8.7.**).

The calculation steps presented in **Subheading 3.9.** have been incorporated into a spreadsheet. Thus, the user simply enters the raw values for the NT, CT, and heterodimer PCR products for each gene into the spreadsheet, and the expression value for the gene in molecules/10^6 β-actin molecules is automatically calculated. Software now in development will automatically enter the peak area values for each NT and CT PCR product into a spread sheet. The spreadsheet will automatically calculate expression value or, if the NT/CT ratio is not in balance, will instruct the robotic liquid handler on how to set up the next experiment.

Acknowledgments

These studies were funded by NCI grants U01 CA 85147 and R24 CA 95806 and the George Isaac Endowment for Cancer Research. Major contributions to establishment of the SEM Center have been made by David A. Weaver. JCW, ELC, KAW, and RJZ have significant equity interest in Gene Express Inc., which produces and markets StaRT-PCR reagents. EAH and RJZ are employees of Gene Express Inc.

References

1. Marshall, E. (1999) Do-it-yourself gene watching. *Science* **286**, 444–447.
2. Bustin, S. A. (2002) Quantification of mRNA using real-time reverse transcription PCR (RT-PCR): trends and problems. *J. Mol. Endocrinol.* **29**, 23–39.
3. Gilliland, G., Perrin, S., Blanchard, K., and Bunn, H. F. (1990) Analysis of cytokine mRNA and DNA: Detection and quantitation by competitive polymerase chain reaction. *Proc. Natl. Acad. Sci. USA* **87**, 2725–2729.
4. Apostolakos, M. J., Schuermann, W. H., Frampton, M. W., Utell, M. J., and Willey, J. C. (1993) Measurement of gene expression by multiplex competitive polymerase chain reaction. *Anal. Biochem.* **213**, 277–284.
5. Willey, J. C., Crawford, E. L., and Jackson, C. M. (1998) Expression measurement of many genes simultaneously by quantitative RT-PCR using standardized mixtures of competitive templates. *Am. J. Respir. Cell Mol. Biol.* **19**, 6–17.
6. Crawford, E. L., Peters, G. J., Noordhuis, P., et al. (2001) Reproducible gene expression measurement among multiple laboratories obtained in a blinded study using standardized RT (StaRT)-PCR. *Mol. Diagn.* **6**, 217–225.
7. Crawford, E. L., Warner, K. A., Khuder, S. A., et al. (2002) Multiplex standardized RT-PCR for expression analysis of many genes in small samples. *Biochem. Biophys. Res. Commun.* **293**, 509–516.
8. Crawford, E. L., Khuder, S. A., Durham, S. J., et al. (2000) Normal bronchial epithelial cell expression of glutathione transferase P1, glutathione transferase M3, and glutathione peroxidase is low in subjects with bronchogenic carcinoma. *Cancer Res.* **60**, 1609–1618.
9. DeMuth, J. P., Jackson, C. M., Weaver, D. A., et al. (1998) The gene expression index c-*myc* x E2F1/p21 is highly predictive of malignant phenotype in human bronchial epithelial cells. *Am. J. Respir. Cell. Mol. Biol.* **19**, 18–24.
10. Mollerup, S., Ryberg, D., Hewer, A., Phillips, D. H., and Haugen, A. (1999) Sex differences in lung CYP1A1 expression and DNA adduct levels among lung cancer patients. *Cancer Res.* **59**, 3317–3320.
11. Rots, M. G., Willey, J. C., Jansen, G., et al. (2000) mRNA expression levels of methotrexate resistance-related proteins in childhood leukemia as determined by a standardized competitive template-based RT-PCR method. *Leukemia* **14**, 2166–2175.

12. Rots, M. G., Pieters, R., Peters, G. J., et al. (1999) Circumvention of methotrexate resistance in childhood leukemia subtypes by rationally designed antifolates. *Blood* **94,** 3121–3128.

13. Allen, J. T., Knight, R. A., Bloor, C. A., and Spiteri, M. A. (1999) Enhanced insulin-like growth factor binding protein-related protein 2 (connective tissue growth factor) expression in patients with idiopathic pulmonary fibrosis and pulmonary sarcoidosis. *Am. J. Respir. Cell. Mol. Biol.* **21,** 693–700.

14. Loitsch, S. M., Kippenberger, S., Dauletbaev, N., Wagner, T. O., and Bargon, J. (1999) Reverse transcription-competitive multiplex PCR improves quantification of mRNA in clinical samples- application to the low abundance CFTR mRNA. *Clin. Chem.* **45,** 619–624.

15. Vondracek, M. T., Weaver, D. A., Sarang, Z., et al. (2002) Transcript profiling of enzymes involved in detoxification of xenobiotics and reactive oxygen in human normal and Simian virus 40 T antigen-immortalized oral keratinocytes. *In. J. Cancer* **99,** 776–782.

16. Ding, C. and Cantor, C. R. (2003) A high-throughput gene expression analysis technique using competetive PCR and matrix-assisted laser desorption ionization time-of-flight MS. *Proc. Natl. Acad. Sci. USA* **100,** 3059–3064.

17. Zhang, J., Day, I. N. M., and Byrne, C. D. (2002) A novel medium-throughput quantitative competitive PCR technology to simultaneously measure mRNA levels from multiple genes. *Nucleic Acids Res.* **30,** e20.

18. Becker-Andre, M. and Hahlbrock, K. (1989) Absolute messenger-RNA quantification using the polymerase chain-reaction (PCR) - a novel-approach by a PCR aided transcript titration assay (patty) nucleic acids research. *Nucleic Acids Res.* **17,** 9437–9446.

19. Wang, A. M., Doyle, M. V., and Mark, D. F. (1989) Quantitation of mRNA by the polymerase chain reaction. *Proc. Natl. Acad. Sci. USA* **86,** 9717–9972.

20. Zhang, J. and Byrne, C.D. (1997) A novel highly reproducible quantitative competitive RT-PCR system. *J. Mol. Biol.* **274,** 338–352.

21. Zhou, N. M., Matthys, P., Polacek, C., et al. (1997) A competitive RT-PCR method for the quantitative analysis of cytokine mRNAs in mouse tissues. *Cytokine* **9,** 212–218.

22. Lyon, E., Millson, A., Lowery, M. C., et al. (2001) Quantification of HER2/neu gene amplification by competitive PCR using fluorescent melting curve analysis. *Clin. Chem.* **47,** 844–851.

23. Hirano, T., Haque, M., and Utiyama, H. (2002) Theoretical and experimental dissection of competitive PCR for accurate quantification of DNA. *Anal. Biochem.* **303,** 57–65.

24. Blaschke, V., Reich, K., Blaschke, S., Zipprich, S., and Neumann, C. (2002) Quantitative RT-PCR: comparing real-time LightCycler technology with quantitative competitive RT-PCR, *Biochemica* **1,** 6–7, www.Roche-Applied-Science.com, http://www.roche-applied-science.com/biochemica/no1_02/PDF/p6.pdf

25. Livak, K. J. and Schmittgen, T. D. (2001) Analysis of related gene expression data using real-time quantitative RT-PCR and the 2 (-Delta Delta C(T)) method. *Methods* **25,** 402–408.

26. Meijerink, J., Mandigers, C., van de Locht, L., et al. (2001) A novel method to compensate for different amplification efficiencies between patient DNA samples in quantitative real-time PCR. *J. Mol. Diagn.* **3,** 55–61.
27. Akane, A., Matsuara, K., Nakamura, H., Takahashi, S., and Kimura, K. (1994) Identification of the heme compound co-purified with deoxyribonucleic acid (DNA) from blood stains, a major inhibitor of polymerase chain reaction (PCR) amplification. *J. Forensic Sci.* **39,** 362–372.
28. Zhu, Y. H., Lee, H. C., and Zhang, L. (2002) An examination of heme action in gene expression: Heme and heme deficiency affect the expression of diverse genes in erythroid K562 and neuronal PC12 cells. *DNA Cell Biol.* **21,** 333–346.
29. Giulietti, A., Overbergh, L., Valckx, D., et al. (2001) An overview of real-time quantitative PCR: applications to quantify cytokine gene expression. *Methods* **25,** 386–401.
30. Heid, C. A., Stevens, J., Livak, K. J., and Williams, P. M. (1996) A novel method for real time quantitative RT-PCR. *Genome Res.* **6,** 986–994.
31. Winer, J., Jung, C. K. S., Shackel, I., and Williams, P. M. Development and validation of real-time quantitative reverse transcriptase-polymerase chain reaction for monitoring gene expression in cardiac myocytes in vitro. *Anal. Biochem.* **270,** 41–49.
32. Bustin, S. A. (2000) Absolute quantification of mRNA using real-time reverse transcription polymerase chain reaction assays. *J. Mol. Endorinol.* **25,** 169–193.
33. Crawford, E. L., Warner, K. A., Weaver, D. A., and Willey, J. C. (2001) Quantitative end-point RT-PCR gene expression measurement using the Agilent 2100 Bioanalyzer and standardized RT-PCR. http://www.chem.agilent.com/temp/rad6A17F/00029012.pdf.
34. Chomczynski, P. and Sacchi, N. (1993) Single-step method of RNA isolation by acid guanidinium thiocyanate-phenol-chloroform extraction. *Anal. Biochem.* **62,** 156–159
35. Willey, J. C., Coy, E. L., Frampton, M. W., et al. (1997) Quantitative RT-PCR measurement of cytochromes p450 1A1, 1B1, and 2B7, microsomal epoxide hydrolase, and NADPH oxidoreductase expression in lung cells of smokers and nonsmokers. *Am. J. Respir. Cell Mol. Biol.* **17,** 114–124.
36. Celi, F. S., Zenilman, M. E., and Shuldiner, A. R. (1993) A rapid and versatile method to synthesize internal standards for competitive PCR. *Nucleic Acids Res.* **21,** 1047.

4

GeneCalling

Transcript Profiling Coupled to a Gene Database Query

Richard A. Shimkets

Summary

We describe the GeneCalling method for the discovery of differentially expressed genes, both known and novel, from any species including useful sequence information to determine the potential function of novel genes captured. The method relies on transcript visualization coupled to a database query to rapidly and quantitatively identify differentially expressed transcripts. The method has been applied to a wide variety of disease models in a variety of species, addressing problems as diverse as identifying novel human cancer gene targets, understanding how drugs and diet affect animal models of disease, and understanding the basis of trait differences in related strains of corn.

Key Words: Bioinformatics, cDNA, disease, GeneCalling, mRNA

1. Introduction

The comprehensive discovery of differences in gene expression among samples is a powerful method of identifying genes associated with diseases, traits, and biological responses to chemicals. Existing methods for expression analysis fall into three general classes: transcript sampling by sequencing *(1–3)*, transcript amplification and imaging *(4–8)* and hybridization-based approaches *(9–13)*. Serial analysis of gene expression (SAGE) *(2)*, a cost-effective transcript-counting technique, is limited by the small amount of sequence information obtained for each gene. Transcript sequencing following subtractive hybridization also identifies differentially expressed genes, but is limited to binary comparisons *(3)*. Transcript imaging approaches such as differential display

From: *Methods in Molecular Biology, Vol. 258: Gene Expression Profiling: Methods and Protocols*
Edited by: R. A. Shimkets © Humana Press Inc., Totowa, NJ

(4), partitioning by type IIS restriction enzymes *(6)*, representational difference analysis (RDA) *(7)*, and amplified fragment length polymorphism (AFLP)
(8) are rapid, and in theory, are comprehensive because they utilize banding patterns that are dependent on gene expression. However, each of these approaches
requires a time-consuming cloning and confirmation process for determination
of the identity of differentially expressed gene fragments.

The development of microarrays has revolutionized the capacity of hybridization techniques *(9–13)* to identify differences in gene expression. Hybridization approaches are rapid and immediately provide the identity of differentially expressed genes of known sequence. However, hybridization methods
are limited by an inability to detect or discover completely novel genes with
no expressed sequence tags (EST) representation, thus making work in most
organisms impossible.

We describe here the GeneCalling® method for the discovery of differentially expressed genes, both known and novel, from any species and with useful sequence information to determine the potential function of novel genes
captured (**Fig. 1**) *(14)*. The method has been applied to a wide variety of disease models in a wide variety of species, addressing problems as diverse as
identifying novel human cancer gene targets *(15,16)*, understanding how drugs
and diet affect animal models of disease *(17,18)*, and understanding the basis
of trait differences in related strains of corn *(19,20)*.

2. Materials

1. Trizol (BRL, Grand Island NY).
2. Bromochloropropane (Molecular Research Center Inc., Cincinnati, OH).
3. DNAse I (Promega, Madison, WI).
4. Dithiothreitol (DTT) (BRL, Grand Island, NY)
5. RNasin (Promega, Madison, WI).
6. OliGreen (Molecular Probes, Eugene, OR).
7. Oligo(dT) magnetic beads (PerSeptive, Cambridge, MA).
8. Superscript II reverse transcriptase (BRL, Grand Island, NY).
9. *E. coli* DNA ligase (BRL, Grand Island, NY).
10. *E. coli* DNA polymerase (BRL, Grand Island, NY).
11. *E. coli* RNase H (BRL, Grand Island, NY).
12. Arctic shrimp alkaline phosphatase (USB, Cleveland, OH).
13. PicoGreen (Molecular Probes, Eugene, OR).
14. Klentaq (Clontech Advantage).
15. PFU (Stratagene, La Jolla, CA).
16. MPG streptavidin beads (CPG).
17. TAMRA- and ROX-tagged molecular size standard (PE-Applied Biosystems, Fos-
    ter City, CA).

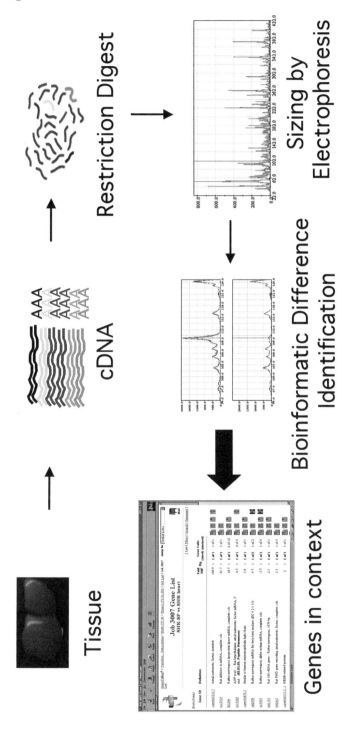

Fig. 1. GeneCalling process diagram.

3. Methods

3.1. GeneCalling Chemistry

1. Isolate total cellular RNA with Trizol using one-tenth volume of bromochloropropane for phase separation.

2. Remove contaminating DNA by treatment with DNAse I in the presence of 0.01 M DTT and 1 unit/μl Rnasin. Following phenol/chloroform extraction, evaluate RNA quality by spectrophotometry and formaldehyde agarose gel electrophoresis, and estimate RNA yield by fluorometry with OliGreen. Prepare Poly-A$^+$ RNA from 100 μg total RNA using oligo(dT) magnetic beads, and quantitate with fluorometry.

3. Prepare first strand cDNA from 1.0 μg of poly(A) + RNA with 200 pmols oligo(dT) 25V (V = A, C, or G) using 400 units of Superscript II reverse transcriptase.

4. Second strand synthesis is performed at 16°C for 2 h following the addition of 10 units of *E. coli* DNA ligase, 40 units of *E. coli* DNA polymerase, and 3.5 units of *E. coli* RNase H. Next, add 5 units of T4 DNA polymerase, and continue incubation at 16°C for 5 min. Treat the reaction with 5 units of arctic shrimp alkaline phosphatase at 37°C for 30 min, and purify cDNA by phenol/chloroform extraction.

5. Estimate the yield of cDNA using fluorometry with PicoGreen.

6. Perform cDNA fragmentation, tagging, and amplification in a three-step process. Achieve fragmentation by restriction enzyme digestions in a 50 μL reaction mix containing 5 units of each restriction enzyme, 1 ng of double-stranded cDNA and 5 μL of the appropriate 10 X buffer restriction endonuclease buffer. Coverage of most mRNAs is achieved by performing 80 separate sets of cDNA fragmentation reactions, each with a different pair of restriction enzymes.

7. Tagging is achieved by ligation of amplification cassettes with ends compatible to the 5' and 3' ends of the cDNA fragments. Incubate the ligation at 16°C for 1 h in 10 mM ATP, 2.5% PEG, 10 units T4 DNA ligase and 1 X ligase buffer.

8. Amplification is achieved by the addition of the following reagents: 2 μL 10 mM dNTP, 5 μL 10 X TB buffer (500 mM Tris-HCl, 160 mM $(NH_4)_2SO_4$, 20 mM $MgCl_2$, pH 9.15), 0.25 μL Klentaq:PFU (16/1), 32.75 μL H_2O. 20 cycles of amplification (30 s at 96°C, 1 min at 57°C, 2 min at 72°C) are followed by 10 min at 72°C.

9. Perform PCR product purification by using MPG streptavidin beads. After washing the beads twice with buffer 1 (3 M NaCl, 10 mM Tris-HCl, 1 mM EDTA, pH 7.5), 20 μL of buffer 1 is mixed with the PCR product for 10 min at room temperature, separated with a magnet, and washed once with buffer 2 (10 mM Tris-HCl, 1 mM EDTA, pH 8.0). Dry the beads and resuspend in 3 μL of buffer 3 (80% (v/v) formamide, 4 mM EDTA, 5% TAMRA- or ROX-tagged molecular size standard.

10. Following denaturation (96°C for 3 min), samples are loaded onto 5% polyacrylamide, 6M urea, 0.5 X TBE ultrathin gels and electrophoresed on a Niagara instrument. The primary components of the Niagara gel electrophoresis system are an interchangeable horizontal ultrathin gel cassette mounted in a platform employing stationary laser excitation and a multicolor CCD imaging system. Each gel cassette is loaded in four cycles of 12 wide (48 lanes total) directly from a 96-well

plate using a robotic arm. The Niagara system has the advantage of high throughput, with separation of fragments between 30 and 450 bases in 45 min. Alternatively, vertical electrophoresis or capillary electrophoresis can be used.

11. PCR products are visualized by virtue of the fluorescent FAM label at the 5' end
of one of the PCR primers, which ensures that all detected fragments have been
digested by both enzymes.

3.2. Gel Interpretation

The output of the electrophoresis instruments are processed using the Java-
based internet-ready Open Genome Initiative (OGI) software suite. First, gel
images are visually checked and tracked. Each lane contains the FAM-labeled
products of a single GeneCalling reaction plus a sizing ladder spanning the
range from 50 to 500 bp. The ladder peaks provide a correlation between camera
frames (collected at 1 Hz) and DNA fragment size in base pairs. After track-
ing, lanes are extracted and the peaks in the sizing ladder are found. Linear
interpolation between the ladder peaks is used to convert the fluorescence traces
from frames to base pairs. A final quality control step checks for low signal-to-
noise, poor peak resolution, missing ladder peaks, and lane-to-lane bleed. Data
that pass all of these criteria are submitted as point-by-point length vs amplitude addresses to an Oracle 8 database.

3.3. Difference Identification

For each restriction-enzyme pair per sample set, calculate a composite trace
by compiling all the individual sample replicates followed by application of a
scaling algorithm for best-fit to normalize the traces of the experimental set vs
that of the control. The scaled traces are then compared on a point-by-point
basis to define areas of amplitude difference that meet the minimum prespeci-
fied threshold for a significant difference. Once a region of difference is identified, the local maximum for the corresponding traces of each set is then identified.
The variance of the difference is determined by

$$\sigma_\Delta^2(j) = \lambda_1(j)^2 \, \sigma_{Total}^2(j:S_1) + \lambda_2(j)^2 \, \sigma_{Total}^2(j:S_2)$$

where $\lambda_1(j)$ and $\lambda_2(j)$ represent scaling factors and $(j:S)$ represents the trace composite values over multiple samples. The probability that the difference is statistically significant is calculated by

$$P(j) = 1 - \int_{-\Delta}^{\Delta} dy \, \frac{1}{\sqrt{2\pi\sigma_\Delta^2}} \exp\left(\frac{-y^2}{2\sigma_\Delta^2}\right)$$

where y is the relative intensity. All difference peaks are stored as unique database addresses in the specified expression difference analysis.

3.4. Gene Identification

cDNA fragments representing differentially expressed genes can be identified by database searching with the six base-pair restriction enzyme recognition sequences at the fragment ends and the exact length of each fragment (determined electrophoretically, subtracting linker length) (*see* **Note 1**). Database searching for genes predicted to have restriction fragments of matching lengths enables the immediate identification of all of the genes whose sequences reside in that database and flags fragments derived from novel genes by virtue of their absence from the database. Given a three-nucleotide size window, database lookup can provide a unique assignment of gene identity. The detection of multiple fragments derived from the same gene that show differential expression of the same directional modulation increases the likelihood that the prediction of the gene identity is correct (**Fig. 2**).

Database lookup can provide a unique assignment of gene identity, and the detection of multiple fragments derived from the same gene that show differential expression of the same directional modulation increases the likelihood that the prediction of the gene identity is correct.

3.5. Gene Confirmation by Oligonucleotide Poisoning

Restriction fragments that map end sequence and length to known genes in the species of interest are used as templates for the design of unlabeled oligonucleotide primers. An unlabeled oligonucleotide designed against one end of the restriction fragment is added in excess to the original reaction, and is reamplified for an additional 15 cycles. This reaction is then electrophoresed and compared to a control reaction reamplified without the unlabeled oligonucleotide to evaluate the selective diminution of the peak of interest.

4. Notes

1. Because the biotin label is necessary for purification and the FAM label is necessary for detection, all detected fragments result from restriction digestion with both enzymes. Typically 96 GeneCalling reactions are performed, each with a separate pair of endonucleases, on triplicate samples.

The principle advantages of GeneCalling include the flexibility to discover known and novel dysregulated genes, the ability to apply this technology to any organism containing tangible RNA, the capturing of the transcript's center, which provides protein-coding information, the ability to sensitively distinguish rare and abundant transcripts, the ability to independently measure transcript abundance multiple independent times in a single experiment, and the ability to comprehensively measure the majority of transcripts in a cell. These characteristics make GeneCalling an attractive system for the drug discovery industry as well as a variety of other molecular biology applications.

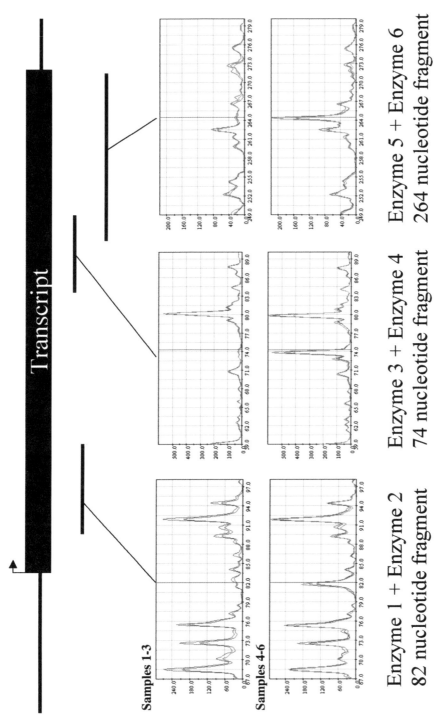

Fig. 2. GeneCalling example data.

References

1. Lee, N. H., Weinstock, K. G., Kirkness, E. F., et al. (1995) Comparative expressed-sequence-tag analysis of differential gene expression profiles in PC-12 cells before and after nerve growth factor treatment. *Proc. Natl. Acad. Sci. USA* **92,** 8303–8307.
2. Velculescu, V., Zhang, L., Vogelstein, B., and Kinzler, K. (1995) Serial analysis of gene expression. *Science* **270,** 484–487.
3. Lee, S. and Tomasetto, C., and Sager, R. (1991) Positive selection of candidate tumor-suppressor genes by subtractive hybridization. *Proc. Natl. Acad. Sci. USA* **88,** 2825–2829.
4. Liang, P. and Pardee, A. B. (1992) Differential display of eukaryotic messenger RNA by means of the polymerase chain reaction. *Science* **257,** 967–970.
5. Ivanova, N. B. and Belyavsky, A. V. (1995) Identification of differentially expressed genes by restriction endonuclease-based gene expression fingerprinting. *Nucleic Acids Res.* **23,** 2954–2958.
6. Kato, K. (1995) Description of the entire mRNA population by a 3' end cDNA fragment generated by class IIS restriction enzymes. *Nucleic Acids Res.* **23,** 3685–3690.
7. Hubank, M. and Schatz, D. G. (1994) Identifying differences in mRNA expression by representational difference analysis of cDNA. *Nucleic Acids Res.* **22,** 5640–5648.
8. Bachem, C. W. B., van der Hoeven, R. S., de Bruijin, S. M., Vreugdenhil, D., Zabeau, M., and Visser, R. G. F. (1996) Visualization of differential gene expression using a novel method of RNA fingerprinting based on AFLP: analysis of gene expression during potato tuber development. *Plant J.* **9,** 745–753.
9. Schena, M., Shalon, D., Davis, R. W., and Brown, P. O. (1995) Quantitative monitoring of gene expression patterns with a complementary DNA microarray. *Science* **270,** 467–470.
10. Lockhart, D. J., Dong, H., Byrne, M. C., et al. (1996) Expression monitoring by hybridization to high-density oligonucleotide arrays. *Nat. Biotechnol.* **14,** 1675–1680.
11. Shalon, D., Smith, S. J., and Brown, P. O. (1996) A DNA microarray system for analyzing complex DNA samples using two-color fluorescent probe hybridization. *Genome Res.* **6,** 639–645.
12. Wodicka, L., Dong, H., Mittmann, M., Ho, M. H., and Lockhart, D. J. (1997) Genome-wide expression monitoring in *Saccharomyces cerevisiae. Nat. Biotechnol.* **15,** 1359–1367.
13. DeRisi, J. L., Iyer, V. R., and Brown, P. O. (1997) Exploring the metabolic and genetic control of gene expression on a genomic scale. *Science* **278,** 680–686.
14. Shimkets R. A., Lowe D. G., Tai J. T., et al. (1999) Gene expression analysis by transcript profiling coupled to a gene database query. *Nat. Biotechnol.* **8,** 798–803.
15. Herrmann, J. L., Rastelli, L., Burgess, C. E., et al. (2001) Implications of onco-genomics for cancer research and clinical oncology. *Cancer J.* **1,** 40–51.
16. Kahn, J., Mehraban, F., Ingle, G., et al. (2000) Gene expression profiling in an in vitro model of angiogenesis. *Am. J. Pathol.* **6,** 1887–1900.
17. Rininger, J. A., DiPippo, V. A., and Gould-Rothberg, B. E. (2000) Differential gene expression technologies for identifying surrogate markers of drug efficacy and toxicity. *Drug Discov. Today* **12,** 560–568.

18. Basson, M. D., Liu, Y. W., Hanly, A. M., Emenaker, N. J., Shenoy, S. G., and Gould Rothberg, B. E. (2000) Identification and comparative analysis of human colonocyte short-chain fatty acid response genes. *J. Gastrointest. Surg.* **5,** 501–512.

19. Bruce, W., Desbons, P., Crasta, O., and Folkerts, O. (2001) Gene expression profiling of two related maize inbred lines with contrasting root-lodging traits. *J. Exp. Bot.* **52(Suppl.),** 459–468.

20. Bruce, W., Folkerts, O., Garnaat, C., Crasta, O., Roth, B., and Bowen, B. (2000) Expression profiling of the maize flavonoid pathway genes controlled by estradiol-inducible transcription factors CRC and P. *Plant Cell* **1,** 65–80.

5

Invader Assay for RNA Quantitation

Marilyn C. Olson, Tsetska Takova, LuAnne Chehak,
Michelle L. Curtis, Sarah M. Olson, and Robert W. Kwiatkowski

Summary

The Invader® assay is a homogeneous, isothermal, signal amplification system for the quantitative detection of nucleic acids. The assay can directly detect either DNA or RNA without target amplification or reverse transcription. It is based on the ability of Cleavase® enzymes to recognize as a substrate and cleave a specific nucleic acid structure generated through the hybridization of two oligonucleotides to the target sequence. The combination of sequence-specific oligonucleotide hybridization and structure-specific enzymatic cleavage results in a highly specific assay well suited for discriminating closely related gene sequences. This includes detection of single nucleotide polymorphisms directly from genomic DNA as well as highly homologous mRNAs in closely related gene families. Because Cleavase® substrate recognition is structure, and not sequence dependent, cleavage and detection can be applied to virtually any DNA or RNA sequence.

Key Words: Cleavase, hybridizationcDNA, Invader, mRNA

1. Introduction

The Invader® assay is a homogenous, isothermal, signal amplification system for the quantitative detection of nucleic acids (*1–3*). The assay can directly detect either DNA or RNA without target amplification or reverse transcription. It is based on the ability of Cleavase® enzymes to recognize as a substrate and cleave a specific nucleic acid structure generated through the association of two oligonucleotides (oligo)s with the target sequence (*4,5*). The combination of sequence-specific oligonucleotide hybridization and structure-specific enzymatic cleavage results in a highly specific assay well suited for discriminating closely related gene sequences. This includes detection of single nucleotide polymorphisms

From: *Methods in Molecular Biology, Vol. 258: Gene Expression Profiling: Methods and Protocols*
Edited by: R. A. Shimkets © Humana Press Inc., Totowa, NJ

A

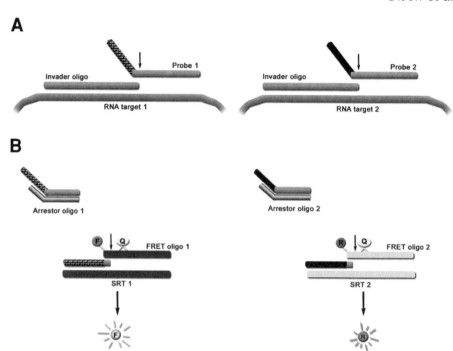

Fig. 1. Schematic representation of the biplex Invader RNA assay. (**A**) Primary Reaction: Probes and Invader Oligos form an invasive structure on the RNA targets. Arrow indicates the cleavage site. (**B**) Secondary Reaction: cleaved 5' flaps (generated in the primary reaction) and the FRET oligos bind to the secondary reaction template (SRT) to form invasive structures recognized by the Cleavase enzyme. Cleavage between the fluorophore (F or R) and the quencher molecule (Q) generates fluorescence signal. The Arrestor oligos sequester the uncleaved probes.

(SNPs) directly from genomic DNA (*1,6,7*) as well as highly homologous mRNAs in closely related gene families (*3,8*). Because Cleavase substrate recognition is structure, and not sequence dependent, cleavage and detection can be applied to virtually any DNA or RNA sequence.

A schematic representation of the Invader RNA Assay is shown in **Fig. 1**. In the primary reaction, the Invader oligo and probe bind specifically to the RNA target and form a one-base overlap, or invasive, structure. The probe consists of a 3' target specific region (TSR) and a 5' flap that is not complementary to the target. The thermostable Cleavase enzyme recognizes the invasive structure formed by the Invader and probe oligos as a substrate and precisely cleaves the 5' flap at the position where the 3' end of the Invader oligo overlaps the probe and target (indicated by the arrow in **Fig. 1**).

The cleavage product therefore includes the 5' flap plus one base of the TSR. The melting temperature (Tm) of the probe TSR is designed to be approx 60°C.

The probe is inherently unstable and "cycles" at the 60°C isothermal reaction temperature, going through multiple rounds of association and dissociation per minute. In contrast, the Invader oligo remains bound to the RNA target. Turnover (association, cleavage, dissociation, and replacement) of the probe, which is present in excess, occurs rapidly. Thus, multiple copies of the probe oligo are cleaved for each copy of the target sequence, without temperature cycling. Typically, 20–30 probes are cleaved per RNA target per minute resulting in signal amplification of approx 2000-fold per target in a 1-h primary reaction *(9)*. The cleavage products (5' flaps) accumulate linearly at a rate proportional to the amount of target in the original sample.

The addition of a secondary reaction provides further signal amplification and a universal detection mechanism. In the secondary reaction, the cleavage product of the primary reaction (the cleaved 5' flap plus one base of the TSR) hybridizes with the Secondary Reaction Template (SRT) and forms a one-base invasive structure with a fluorescence resonance energy transfer (FRET) oligo. Enzymatic cleavage of the FRET oligo separates a fluorophore (F) from a quencher molecule (Q) to generate signal. Multiple FRET oligos can be cleaved for each 5' flap generated in the primary reaction resulting in an overall amplification of fluorescence signal of approx 10^6-fold. The sequence and length of the 5' flap is designed so that it remains bound to the SRT, which is required for efficient signal generation. However, uncleaved probes carried over from the primary reaction can also bind stably to the SRT and inhibit signal generation in the secondary reaction by competing with the cleaved 5' flaps. Adding an Arrestor oligo to the secondary reaction reduces competitive inhibition. The Arrestor oligo is complementary to the probe TSR and a portion of the 5' flap and is therefore able to sequester the uncleaved probe. This prevents the uncleaved probes, but not the 5' flaps, from binding to the SRT during the secondary reaction. The 5' flap, SRT and FRET oligo are not target-specific therefore the same detection oligos can be used for many different genes which simplifies assay design and lowers production costs.

The biplex Invader RNA assay format enables simultaneous detection of two different genes within the same sample *(3)*. This is accomplished by using two unique 5' flaps on the target specific probes that differ in sequence but have similar Tm so that both 5' flaps can bind to their complementary SRTs at the 60°C reaction temperature. Typically, one 5' flap sequence is used for detection of genes of interest and the other 5' flap sequence for housekeeping genes. This enables assays for any one of several different housekeeping genes to be readily combined with an mRNA assay for added flexibility. Two different SRT and FRET oligos are used in the biplex assay. The FRET oligos contain a Z28 quencher molecule (Epoch Biosciences, WA) and two spectrally distinct fluoro-phores FAM (F) and Redmond Red™ (R) (Epoch Biosciences). The biplex format permits normalization to an internal control (housekeeping gene).

2. Materials

2.1. Sample Preparation

1. Total RNA can be isolated from cells or tissues using standard reagents such as TRIzol® (Invitrogen, Carlsbad CA, cat. no. 15596–026) or RNeasy® (Qiagen, Valencia, CA, cat. no. 74124). Store total RNA samples at −70°C.
2. Cell lysates are prepared using a lysis buffer containing 20 mM Tris-HCl, pH 8, 5 mM MgCl$_2$,0.5% NP40, 20 ng/µL of tRNA.
3. tRNA carrier at 20 ng/µL (Sigma, cat. no. R-5636) is used as a no target control and for preparation of in vitro transcript dilutions.
4. PBS, no MgCl$_2$/no CaCl$_2$ (for cell lysate preparation only).
5. RNase-free (DEPC-treated) H$_2$O.

2.2. Invader RNA Assay Reagents

2.2.1. Oligonucleotides

1. Gene-specific oligos: The Probe, Invader oligo, Arrestor and Stacker (optional). Assays are available from Third Wave Technologies for a number of genes. All predeveloped assays contain primary oligo mixes and secondary detection oligos along with a corresponding RNA standard (in vitro transcript RNA). The target specific region of the probe is designed to maintain specificity through appropriate site selection that is dependent on the target of interest. Optimum signal generation at a predetermined reaction temperature of 60°C is achieved by adjusting the length of the target-specific region (TSR) so that the Tm is close to 60°C. Invader Creator™ software (Third Wave Technologies) is used to make the Invader assay-specific adjustments to nearest neighbor Tm predictions (*10,11*). The 5'-flap sequence is chosen for compatibility with predeveloped secondary detection components.
2. Detection Oligos: Secondary Reaction Templates (SRT) and FRET Oligos. Detection oligos are available from Third Wave Technologies for use with the standard 5' flaps (*see* **Subheading 3.1.2.**) FAM (cat. no. 91–242) and Red (cat. no. 91–241).
   All diluted oligos should be stored at −20°C.

2.2.2. Generic Reagents

Generic Reagents kits optimized for the Invader RNA Assay (Third Wave Technologies, cat. no. 91–080) contain the following components:

1. 40 ng/µL Cleavase IX Enzyme.
2. RNA primary buffer: 25 mM MOPS, pH.7.5, 250 mM KCl, 0.125% Tween-20, 0.125% NP-40, 31.25 mM MgSO$_4$, 10% PEG.
3. RNA secondary buffer: 87.5 mM MgSO$_4$.

4. tRNA carrier: 20 ng/µL.
5. $T_{10}e_{0.1}$ buffer: 10 mM Tris-HCl, pH 8, 0.1mM EDTA.
6. 10X Cell lysis buffer: 200 mM Tris-HCl, pH 7.5, 50 mM MgCl$_2$ 200 µg/mL tRNA, 5% NP-40.

Generic reagents should be stored at −20°C.

Reagents required but not provided in the kit include RNase-free mineral oil (Sigma, cat. no. M-5904) or Clear Chill-out™ liquid wax (MJ Research, cat. no. CHO-1411) used for preventing reagent evaporation during incubation.

2.3. Equipment and Disposables

1. Fluorescence plate reader with filters that accommodate the following wavelength and bandwidth properties:
   FAM Dye - Excitation 485 nm/20nm and Emission 530 nm/25nm
   Redmond Red™ Dye - Excitation 560 nm/20nm and Emission 620 nm/40nm
2. Thermal cycler or oven for 60°C incubation (or 75°C for cell lysate preparation)
3. 96-well polypropylene skirted microplate (MJ Research, cat. no. MSP-9601/natural).

3. Methods

3.1. Invader RNA Assay Design

3.1.1. Determining the Cleavage Site on the Target RNA

Invader RNA Assays can be designed to be highly specific. To do this, the RNA sequence must be analyzed prior to assay design to determine whether homologous sequences exist. Sequence alignments between related RNAs identify nonhomologous regions for positioning the cleavage site. A single base difference is sufficient for discrimination, however, locating regions where multiple nonhomologous bases exist (especially in the probe region) can maximize specificity. The following procedure is used when designing assays for closely related RNAs:

1. Identify any homologous gene sequences using NCBI Blast. http://www.ncbi.nlm.nih.gov/BLAST
2. If homologous sequences exist, use an alignment program such as the Megalign module of the DNAStar Sequence Analysis Package (DNAStar, Madison, WI) to locate sites of discrimination.
3. Design Invader and probe oligo sets so that at least probe position 1 (cleavage site), and preferably position 2 or -1 are located at a nonhomologous site (*see* **Fig. 2**).
4. Verify specificity of design by blasting the sequence of the region covered by the Invader Assay oligonucleotides.

Invader RNA Assays may also be designed to eliminate cross reactivity with genomic DNA. The Invader and Probe oligos can be targeted to span splice junctions so that the invasive structure required for cleavage is created only on

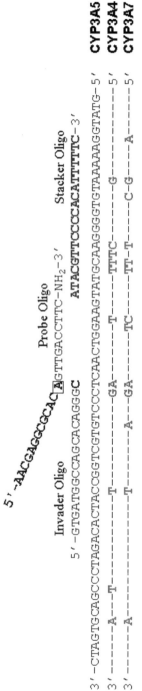

Fig. 2. Discrimination of CYP3A5 mRNA. CYP3A5 Invader RNA assay design and sequence alignment of the homologous 3A4, 3A5, and 3A7 mRNAs (deviations shown on bottom two rows). The sequences of the Invader, Probe and Stacker oligos are indicated above the alignment. The boxed base in the probe designates position 1. Positions −1 and +2 refer to bases on the left and right of the boxed base respectively.

mature mRNA but is not formed on unspliced genomic DNA. Splice junctions are typically listed in the GenBank report (intron/exon sites), but may also be identified by aligning the mRNA and gene sequences. Assay oligo sets are designed with the cleavage site as close to the splice junction as possible. If introns do not exist, cross-reactivity with genomic DNA is avoided through reaction conditions. Specifically, the optimum temperature for detection of any sequence differs on a DNA or RNA target. The lack of a denaturation step in the RNA assay also limits the signal from duplex DNA targets. We have demonstrated that the combination of these factors is sufficient to avoid cross-reactivity between RNA and genomic DNA. Finally, the RNA preparation method can be adapted to eliminate or reduce the amount of DNA contamination.

Another consideration in the selection of the cleavage sites is the accessibility of the target site for hybridization of the Invader assay oligonucleotides. Secondary and tertiary structures characteristic of RNA render much of the sequence inaccessible for hybridization in solution. Because success of the Invader RNA assay depends upon rapid cycling of the signal oligonucleotide probes, we have devised strategies to identify accessible sites on RNA.

The RNAstructure software predicts RNA secondary structure. It is available on the Turner Lab Homepage http://rna.chem.rochester.edu/RNAstructure. html. The Oligo Walk module of RNAstructure selects sites that are more likely to be accessible for oligonucleotide binding *(12)*. Oligo walk uses a set of thermodynamic parameters for RNA, DNA, and their hybrids in an algorithm that relies on mfold for RNA secondary structure prediction. OligoWalk analysis is performed with a 10 base oligonucleotide to resemble the average length of the target specific region of the probe. The affinity of the oligomer to its target is expressed as an overall Gibbs free energy change of a self-structured oligomer and of a target associating into an oligomer -target complex. The lowest negative values generally indicate the most favorable sites for oligonucleotides to bind. The probe (especially the 3' end) is designed to hybridize to these favorable sites. The most inaccessible regions have positive binding energy values and generally are poor sites for assay probe design

Another approach is to experimentally determine accessible sites using the Reverse Transcriptase-Random Oligonucleotide Libraries (RT-ROL) *(13)*. This technique was applied to several different mRNAs. In each case, only a limited number of "accessible" sites were identified (between 5 and 15 on each mRNA). We have observed that Invader assays designed to the identified accessible regions were more sensitive than standard assays. For instance, using this method we have developed an assay that can detect less than 1000 copies of HIV viral RNA *(3)* whereas the standard RNA assay limit of detection is typically 6000 copies. However, the RT-ROL method is more laborious and is only used in cases where high sensitivity is critical.

   The sensitivity of the Invader assay is improved by including a stacking oligo that may create a more accessible region on the RNA target. This oligonucleotide binds to the RNA target and is designed to coaxially stack *(14)* with the 3' end of the probe as shown in **Fig. 2**. The assay performance can be improved further by incorporating 2'-O Me bases into the stacker oligo particularly at the 5' end. Because the stacking interaction increases oligo stability, the probe can be shortened, reducing the probability of deleterious inter- and intramolecular structures interfering with signal generation.

3.1.2. Invader Assay Oligonucleotide Designs

1. Invader oligo design: The Invader oligo is designed so that the Tm is approx78°C or 15°–20°C higher than the Tm of the TSR of the probe. This increases the probability of generating cleavable struture each time the probe cycles on and off the target. The last base at the 3' end of the Invader oligo that overlaps the probe and target, does not need to match the target. In fact, the cleavage rate is typically enhanced by an Invader oligo with a mismatched 3' base. The relative cleavage efficiencies of 3' mismatches have been experimentally determined. Preferred 3' mismatch bases are automatically incorporated into Invader oligos when using the Invader Creator™ software (Third Wave Technologies). The 3' mismatch also permits the use of a universal detection oligos since the 3' end of the cleaved flap (one nucleotide of the probe TSR) does not need to match the secondary reaction template. The bases immediately upstream of the 3'-end must hybridize to the target in order to stabilize the invasion and direct cleavage of the probe.

2. Probe oligo design: The probe oligo consists of two regions; a 3' TSR and a 5' flap that is not complementary to the target. The probe TSR is typically designed so that the Tm is approx 60°C because both the primary and secondary Invader reactions are optimized to perform at this temperature. Assays have been designed to primary reaction temperatures ranging from 50 to 68°C but these assays are not isothermal when using the standard 5' flaps and detection oligos. The actual optimum primary reaction temperature can be determined for each oligo set by testing performance at varying temperatures in a gradient thermal cycler. For any given design, peak performance is observed over a 2–4 degree range. Theoretically, the optimum temperature can be shifted in either direction by lengthening or shortening the TSR of the primary probe. However, a minimum length of nine bases (exclusive of the flap) is required for proper substrate recognition, and lengthening the oligonucleotide increases the risk of forming inter- or intramolecular interactions that can negatively impact performance. Probes are blocked at the 3' end of the oligo with an amine group to prevent possible background signal through hybridization with the SRT, but this may be not necessary for all designs.

   The 5' flap of the probe oligo can vary from 1 to 15 nucleotides in length as long as the sequence does not form stable inter or intramolecular structures. Standard 5' flap sequences have been optimized for optimal performance at 60°C. Oligos containing the following 5' flap sequences are used with the generic detection oligos available from Third Wave Technologies: FAM dye, 5'-AACGAGGCGCAC-3' and for the Redmond Red dye, 5'-CCGCCGAGATCAC-3'.

3. Stacker oligo design: The stacker oligo is designed to stably bind to the RNA target and coaxially stack *(14)* with the 3' end of the probe, thus increasing the probe Tm. Therefore, designs that incorporate a stacker oligo allow shorter probes to effectively cycle at 60°C. Assay performance is improved by incorporating 2'*O*-methyl bases into the stacker oligo particularly when 3–5 bases at the 5' end are modified. The 2'*O*-Me bases also increase the Tm of the oligo (approx 0.5–0.8 degrees/base) when hybridized to a RNA target so shorter oligos remain bound at the 60°C reaction temperature. We routinely incorporate 2'*O*-Me bases in the entire stacker oligo sequence to ensure stable hybridization to the RNA target and to standardize designs. The use of stacker oligos has been shown to improve assay sensitivity but may not be necessary when designing to highly expressed genes such as housekeeping genes.

4. Arrestor oligo design: The Arrestor oligo is used to functionally, but not physically remove the probe from the secondary reaction. Its effects can include both lower background and increased signal. It is designed to be complementary to the probe TSR and extend six bases into the 5' flap. The use of 2'-*O*-methyl bases renders the probe/arrestor complex resistant to Cleavase enzyme activity.

5. Secondary Reaction Templates and FRET oligos. The secondary reaction template is designed to hybridize to both the cleaved 5' flap and FRET oligo. FRET oligos contain either a FAM or Redmond Red™ (Epoch Biosciences) fluorophores and a Z28 dark quencher molecule (Epoch Biosciences). The following SRT sequences are used with the 5' flap sequences mentioned above:
   FAM dye detection, 5'-CCAGGAAGCAAGTGGTGCGCCTCG<u>UUU</u>-3'
   Red dye detection, 5'-CGCAGTGAGAATGAGGTGATCTCGGC<u>GGU</u>-3'
   The underlined bases indicate 2'*O*-methylated nucleotides. The following FRET sequences are used:
   FAM- 5'-CAC(Z28)TGCTTCGTGG-3'
   Red dye - 5'-CTC(Z28)TTCTCAGTGCG-3'

3.1.3. Oligonucleotide Purification and Preparation

Oligonucleotides should be diluted and stored in $T_{10}e_{0.1}$ (10mM Tris-HCl, 0.1mM EDTA, pH 8.0). Mix oligonucleotide stocks prior to dilution and quantization of all oligos. We recommend vortexing the oligo solution followed by brief centrifugation. Quantitate oligos by determining the absorbance at 260 nm. **Table 1** describes the oligonucleotide purification methods and concentrations commonly used in the Invader assay. The probe and FRET oligos should be purified by anion exchange high-performance liquid chromatography (HPLC) because products of incomplete synthesis can cause nonspecific background signal in the Invader assay. HPLC purification of the Invader oligo and stacker oligo is not essential. These oligos can be purifed by NAP desalt, however, signal may be slightly reduced.

Table 1
Invader RNA Assay Oligonucleotide Purification and Reaction Concentrations

Oligo Type	Purification	Working Stock Concentration	Reaction Concentration
Probe	Anion exchange HPLC/C18 desalt	40 μ*M*	10 μ*M*[a]
Invader oligo	Anion exchange HPLC/NAP desalt	20 μ*M*	5 μ*M*[a]
Stacker oligo	Anion exchange HPLC/NAP desalt	12 μ*M*	3 μ*M*[a]
Arrestor oligo	NAP desalt Anion exchange	26.7 μ*M*	2.67 μ*M*[b]
Secondary Reaction Template	HPLC/NAP desalt	1.0 μ*M*	0.1 μ*M*[b]
FRET oligo	Anion exchange HPLC/NAP desalt	6.7 μ*M*	0.67 μ*M*[b]

[a]Final concentrations of primary reaction oligos (Probe, Invader and Stacker) in a 10 μL reaction volume.

[b]Final concentrations of secondary reaction oligos (Arrestor, secondary reaction template and FRET) in a 15 μL (final) reaction volume.

3.2. Sample Preparation

3.2.1. Total RNA Preparation

1. Prepare total RNA from cells or tissues according to manufacturer's instructions.
2. Dilute total RNA samples with RNase-free dH$_2$O. We typically use 50–100 ng of total RNA per reaction but this can vary depending on expression level of the gene. A preliminary experiment is recommended to determine the amount of total RNA (1–100 ng) that generates signal in the linear quantitation range of the assay. High total RNA concentrations (>500 ng/reaction) can inhibit the Invader Assay.

3.2.2. Cell Lysate Preparation

This method is used for adherent cells cultured in 96-well tissue culture plates (10,000–40,000 cells per well).

1. Prepare 1X Cell lysis buffer.
2. Remove culture medium without disturbing the cell monolayer.
3. Wash the cells once with 200 μL of PBS (no MgCl$_2$/no CaCl$_2$). Blot off excess solution because residual PBS can inhibit the assay.
4. Add 40 μL of 1X Cell Lysis Buffer per well. Lyse cells at room temperature for 3–5 min.
5. Transfer 25 μL of each lysate sample to a polypropylene microplate.

Table 2
Invader RNA Assay Primary Reaction Mix
Preparation for Single and Biplex Assay Formats

Reaction Components	1X Volume
Single Assay Format	
RNA Primary Buffer 1	4.0 µL
Primary Oligos (Gene 1)	0.25 µL
$T_{10}e_{0.1}$ Buffer	0.25 µL
Cleavase® IX enzyme	0.5 µL
Total Mix Volume	5.0 µL
Biplex Assay Format	
RNA Primary Buffer 1	4.0 µL
Primary Oligos (Gene 1)	0.25 µL
Primary Oligos (Gene 2)	0.25 µL
Cleavase® IX enzyme	0.5 µL
Total Mix Volume	5.0 µL

6. Overlay lysate samples with 10 µL of Chill-out™(liquid wax or mineral oil (not necessary if using a heated-lid thermal cycler).
7. Cover microplate with well tape. Immediately heat lysates at 75°C for 15 min in a thermal cycler or oven to inactivate cellular nucleases.
8. After the heat inactivation step, add the lysate samples directly to the primary reaction or immediately store at −70°C. Long term stability has not been established and may differ depending on the gene or cell type.

3.2.3. RNA Standard Preparation

The RNA standards or positive controls used in the Invader RNA assays are in vitro transcripts with known concentrations. Serial dilutions of in vitro transcripts are used to generate a standard curve and determine the dynamic range and detection limit of a specific Invader assay design *(15)*. The standard curve is used to accurately quantify specific RNA levels in either total RNA or cell lysate samples.

3.3. Invader RNA Assay

1. Prepare samples and RNA standard dilutions. Example dilution series can be found in the Invader RNA assay product information sheets *(15)*.
2. Prepare primary reaction mix for either the signal or biplex assay format (*see* **Table 2**). To calculate the volumes of reaction components needed for the assay, multiply the number of reactions by 1.25.
3. Mix well and add 5 µL of primary reaction mix to each well of the polypropylene microplate.

4. Add 5 µL of controls or samples and mix by pipeting up and down once or twice. A no target control should be included to determine background signal.
5. Overlay each reaction with 10 µL of Chill-out™ liquid wax or mineral oil.
6. Incubate the microplate for 90 min at 60°C.
7. Prepare secondary reaction for either a single or biplex reaction format (*see* **Table 3**). Calculate the volumes required by multiplying the number of reactions by 1.25.
8. Add 5 µL of secondary reaction mix per well below the Chill-out liquid wax or mineral oil layer using a multichannel pipet. Mix by pipeting up and down once or twice.
9. Incubate the microplate for 60 or 90 min at 60°C.
10. Directly read the plate in a fluorescence plate reader (FAM dye: Ex. 485/20 nm, Em. 530/25 nm, Redmond Red Dye: Ex. 560/20 nm, Em. 620/40 nm). Optimal gain settings can vary between instruments. Adjust the gain as needed to give the best signal/background ratio (sample raw signal divided by the No Target Control signal). The probe height of the fluorescence plate reader may need to be adjusted and a new plate definition map should be created for the microplate (consult the manufacturer's instructions).
11. If the plate cannot be read soon after the secondary incubation is completed, the reaction can be stopped by adding 10 µL of 10 m*M* Tris-HCl, 100 m*M* EDTA to each well. After stopping the reaction, cover the plate in the dark at room temperature for up to 24 h.

3.4. Data Analysis

1. Import the microplate data into Microsoft® Excel or other data analysis program. Determine the average values for the controls and samples (average signal) and calculate the standard deviation (SD) and % coefficient of variance [% CV = (SD/ average signal) × 100].
2. To determine signal/background, divide the average positive control or unknown sample signal by the average no target control signal.
3. To determine net signal, subtract the average no target control signal from the average positive control or unknown sample signal. Generate a standard curve with the positive control net signal values using an appropriate curve fit equation. The polynomial equation is used to fit the data for the samples to the standard curve. The quadratic equation will determine the quantity (x) of a sample. The accuracy of the standard curve can be verified by back-calculating the level of each positive control using the net signal values and the standard curve equation. An example of the data analysis is shown in **Fig. 3**.
4. Calculate RNA levels in unknown samples by using the standard curve equation derived in step 3 and each sample's net signal value. Differences in RNA levels can be determined using appropriate statistical analysis, such as the 95% confidence intervals (95% CI) or *t*-test. The limit of detection (LOD) for a given assay typically corresponds to a Signal/Background value ≥ 1.15 and the *t*-test from the no target control of less than 0.05. Absolute quantitation requires a standard curve. Relative quantitation does not require a standard curve but can be determined from

	RNA Standard Curve Attomoles of IL-8 in vitro transcript							Cell Lysates	
								Uninduced	Induced
	0	0.02	0.04	0.08	0.31	1.25	2.5	0	+
Average	35	43	53	73	180	533	849	37	483
SD	0.00	0.58	1.15	1.15	4.00	8.02	14.64	0.58	28.02
CV	0%	1%	2%	2%	2%	2%	2%	2%	6%
Signal/Bkgd		1.2	1.5	2.1	5.1	15.2	24.2	1.0	13.8
Net Signal		8	18	38	145	498	814	2	448
T-test (0)		0.000	0.000	0.000	0.000	0.000	0.000	0.004	0.000
Calculated amole		0.02	0.04	0.08	0.32	1.25	2.50	0.00	1.09
Copies/cell								ND	523

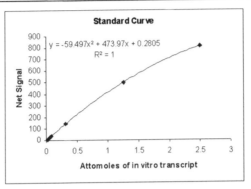

Fig. 3. Quantitation of IL8 mRNA in cell lysate samples. Human IL8 mRNA was quantitated in MG-63 (American Type Culture Collection, Manassas, VA) cell lysate samples by generating a standard curve with in vitro transcript RNA. Lysates from 1250 cells were added to the Invader assay ($n = 3$ for each in vitro transcript RNA standard, $n = 8$ for cell lysate samples).

Table 3
Invader RNA Assay Secondary Reaction Mix
Preparation for Single and Biplex Assay Formats

Reaction Components	1X Volume
Single Assay Format	
RNA Secondary Buffer 1	2.0 μL
Secondary Oligos (Gene 1)	1.5 μL
$T_{10}e_{0.1}$ Buffer	1.5 μL
Total Mix Volume	5.0 μL
Biplex Assay Format	
RNA Secondary Buffer 1	2.0 μL
Secondary Oligos (Gene 1)	1.5 μL
Secondary Oligos (Gene 2)	1.5 μL
Total Mix Volume	5.0 μL

Table 4
Trouble Shooting Guide

Problem	Cause	Solution
Low Signal Generation/ No Signal Generation	• Fluorescence plate reader was not correctly set up.	• Check that the appropriate excitation and emission filters are in place and the instrument is set to read from the top of the plate. • Adjust the gain setting for the best signal/noise ratio. • Probe height may need to be adjusted and a new plate definition should be created according to the manufacturer's instructions.
	• Potential RNase contamination of the samples and reagents.	• Always wear gloves when handling reaction components and use RNase-free solutions and equipment.
	• Oligonucleotides or targets were diluted improperly. • Oligos were diluted in 10 mM Tris-HCl, 1 mM EDTA rather than 10 mM Tris-HCl, 0.1 mM EDTA. • The Arrestor, SRT or FRET oligo was not added.	• If a dilution error is suspected, repeat dilution. • Use only 10 mM Tris-HCl, 0.1 mM EDTA to dilute oligonucleotides. • Repeat reactions with the three essential oligos added in the secondary reaction mix.
	• Incorrect detection oligos were added.	• Repeat run using correct components (i.e., SRT needs to bind with the cleaved 5' flap and FRET oligo).
High Background	• Target added to the No Target Control. • The Probe or FRET oligo was not purified as recommended.	• Check plate layout and repeat run. • Purify Probe and FRET oligo using anion exchange HPLC.

Problem	Possible Cause	Solution
Assay Not Sensitive enough	• Incubation time was reduced.	• Incubate both the primary and secondary reactions for the recommended times. The secondary reaction time can be increased provided that the background is low.
	• Design does not include stacker oligo.	• Redesign with stacker oligo.
	• The primary reaction has an optimum temperature other than 60°C.	• Verify primary reaction temperature by testing Invader reactions, including negative controls and 1–2 moles of target, at 60 +/–5°C. If the reaction peak is below 58°C, increase the length of the probe by one base. If the reaction peak is above 62°C, decrease the length of the probe by one base. Alternatively, the primary reaction can be performed at the optimal primary reaction temperature with a 60°C secondary reaction incubation.
High variation between replicate samples	• Incomplete mixing.	• Thoroughly mix all reagents before dispensing into reaction plate. The secondary reaction mix should be added beneath the overlay.
	• Pipeting error.	• When using a multi-channel pipet, visually inspect tips when aspirating solution to ensure that reagent volumes are equal in all channels.
	• Reaction evaporation.	• Overlay the reactions with Chill-out liquid wax or RNase-free mineral oil.
Signal inhibition from Total RNA or Cell lysate samples	• Total RNA is contaminated with genomic DNA.	• Use a RNA isolation method that minimizes the presence of genomic DNA.
	• Too much total RNA was added to the assay.	• Add 0.1 to 200 ng of total RNA per reaction depending on expression level of the gene. Do not add more than 500 ng of total RNA.
	• Cell lysate preparation contained residual PBS (>5 μL).	• Remove PBS by gently blotting the tissue culture plate on absorbent paper or thorough aspiration.
	• Cells were washed with PBS that contained $CaCl_2$ and $MgCl_2$.	• Wash cells with PBS that does not contain $CaCl_2$ and $MgCl_2$.

the sample net signal values that fall within the linear range of the assay. For relative quantitation, the no target control is required to determine background signal. Additionally, sample signals can be normalized to an invariant housekeeping gene signal using the biplex format for the Invader RNA assay.

4. Notes

1. Use RNase-free disposables and reagents for sample and reaction preparation.
2. The dynamic range of the assay is typically limited to 2–3 logs when using an endpoint read method. Varying the secondary reaction time, sample concentration or using a real-time fluorescence plate reader can extend the dynamic range of the assay.
3. Agarose gel electrophoresis and ethidium bromide staining can assess the purity and integrity of the RNA sample. Genomic DNA can inhibit signal generation if present at high levels and lead to inaccurate RNA quantitation if using A_{260} measurements.
4. High background signal can be caused by unpurified probes and FRET oligos. Anion exchange high-performance liquid chromatography is the recommend purification for these oligos. Although we routinely synthesize Invader Assay oligonucleotides at Third Wave Technologies, oligos have been synthesized by commercial suppliers including Qiagen Operon (CA) and BioSearch Technologies Inc. (CA).
5. When preparing cell lysates, be sure to remove residual PBS before lysing the cells because PBS can inhibit the assay. If the lysate samples are generating unusually high signal across the entire plate, the cellular nucleases may not be heat inactivated. Make sure lysates are heated at 75°C for at least 15 min.
6. In vitro transcript dilutions should be stored in tRNA carrier 20 ng/μL (Sigma, cat. no. R-5636) at −20°C or −70°C (for long-term storage).

References

1. de Arruda, M., Lyamichev, V. I., Eis, P. S., et al. (2002) Invader technology for DNA and RNA analysis: principles and applications. *Expert Rev. Mol. Diagn.* **2,** 487–496.
2. Hall, J. G., Eis, P. S., Law, S. M., et al. (2000) Sensitive detection of DNA polymorphisms by the serial invasive signal amplification reaction. *Proc. Natl. Acad. Sci. USA* **97,** 8272–8277.
3. Eis, P. S., Olson, M. C., Takova, T., et al. (2001) *Nat. Biotechnol.* **19,** 673–676.
4. Lyamichev, V., Brow, M. A., and Dahlberg, J. E. (1993) Structure-specific endonucleolytic cleavage of nucleic acids by eubacterial DNA polymerases. *Science* **260,** 778–783.
5. Ma, W., Kaiser, M. W., Lyamichev, N., et al. (2000) RNA template-dependent 5' nuclease activity of Thermus aquaticus and Thermus thermophilus DNA polymerases. *J. Biol. Chem.* **275,** 24693–24700.
6. Nagano, M., Yamashita, S., Hirano, K., et al. (2002) Two novel missense mutations in the CETP gene in Japanese hyperalphalipoproteinemic subjects: high-throughput assay by Invader assay. *J. Lipid Res.* **43,** 1011–1018.

7. Lyamichev, V. I. and Neri, B. P. (2003) Invader assay for SNP genotyping. Single nucleotide polymorphisms (Kwok, P., ed.). Humana, Totowa, NJ, 229–240.
8. Burczynski, M. E., McMillian, M., Parker, J. B., et al. (2001) Cytochrome P450 induction in rat hepatocytes assessed by quantitative real-time reverse transcription polymerase chain reaction and the RNA invasive cleavage assay. *Drug Metab. Dispos.* **29,** 1243–1250.
9. Lyamichev, V. I., Kaiser, M. W., Lyamicheva, N. E., et al. (2000) Experimental and theoretical analysis of the invasive signal amplification reaction. *Biochemistry* **39,** 9523–9532.
10. Allawi, H. T. and SantaLucia, J. Jr. (1997) Thermodynamics and NMR of internal G.T mismatches in DNA. *Biochemistry* **36,** 10581–10594.
11. Gray, D. M. (1997) Derivation of nearest-neighbor properties from data on nucleic acid oligomers. I. Simple sets of independent sequences and the influence of absent nearest neighbors. *Biopolymers* **42,** 783–793.
12. Mathews, D. H., Burkard, M. E., Freier, S. M., Wyatt, J. R., and Turner, D. H. (1999) Predicting oligonucleotide affinity to nucleic acid targets. *RNA* **5,** 1458–1469.
13. Allawi, H. T., Dong, F., Ip, H. S., Neri, B. P., and Lymichev, V. I. (2001) Mapping of RNA accesible sites by extension of random oligonucleotide libraries with reverse transcriptase. *RNA* **7,** 314–327.
14. Lane, M. J. (1997) The thermodynamic advantage of DNA oligonucleotide stacking hybridization reactions: energetics of a DNA nick. *Nucleic Acids Res.* **25,** 611–617.
15. Invader RNA Assay product information sheet for the human IL8 oligonucleotides and control kit, http://www.twt.com/files/15795.lbl.pdf.

6

Monitoring Eukaryotic Gene Expression Using Oligonucleotide Microarrays

Jennifer Lescallett, Marina E. Chicurel,
Robert Lipshutz, and Dennise D. Dalma-Weiszhausz

Summary

An increasing number of biological and medical research questions depend on obtaining global views of gene expression. In this chapter, we will describe how oligonucleotide microarrays have been used to accomplish this goal. In particular, we will focus on the use of GeneChip arrays®, which provide high levels of reproducibility, sensitivity, and specificity. Target preparation, hybridization, washing, signal detection, and data analysis will be described in detail. Additionally, we will discuss options for facilitating data sharing, including the creation of databases, and the use of internet tools that help users place their results in the context of data from public and proprietary databases.

There is so much interest and innovation in the field of genomics that protocols are constantly evolving. This chapter should be used as a genomic profiling guide only. We urge readers to consult www.affymetrix.com for the most current products and protocols.

Key Words: High-density oligonucleotide microarray, DNA microarray, gene expression, expression profiling, genomics

1. Introduction

The analysis of gene expression is key to addressing a wide variety of medical and biological research questions, including the dissection of basic biological processes, the classification of disease, and the identification of new drug targets. Until recently, comparing expression levels across different tissues or cells was restricted to monitoring a few genes at a time. Using DNA microarrays, however, it is possible to monitor the activities of thousands of genes at once *(1)*.

Global analyses of gene expression can be useful for obtaining in-depth views of cell function. It is estimated, for example, that between 0.2 and 10% of all

From: *Methods in Molecular Biology, Vol. 258: Gene Expression Profiling: Methods and Protocols*
Edited by: R. A. Shimkets © Humana Press Inc., Totowa, NJ

transcripts in a typical mammalian cell are differentially expressed between cancer and normal tissues *(2)*. Whole-genome analyses are also useful because they provide a powerful tool to search through the activities of thousands of genes and identify key players *(3,4)*. In addition, large-scale analyses of expression allow investigators to generate robust classifiers of disease that can outperform traditional, single-marker tests *(5,6)*. Moreover, these analyses frequently yield information that extend beyond the study's original aims. A study designed to identify expression patterns that correlate with a clinical outcome, for example, may also generate insights into the disorder's basic biology, as well as identify candidate drug targets *(5–7)*.

In this chapter, we describe the use of GeneChip® probe arrays, oligonucleotide microarrays that allow global analyses of gene expression with a high degree of reproducibility, sensitivity, and specificity *(8)*. Unlike other microarrays, GeneChip probe arrays track real and stray hybridization signals in a probe-specific manner, enabling accurate detection and quantitation of low-abundance transcripts. In addition, the probes can be designed to distinguish between homologous transcripts that are up to 90% identical *(9)*. The design and manufacture of GeneChip probe arrays is highly stereotyped and consistent, ensuring a high degree of reproducibility between experiments *(10)*. This reproducibility allows the comparison of one control sample to many experimental samples, or several controls to many experimental samples.

In this chapter, we also present practical guidelines for optimizing the capabilities of GeneChip probe arrays. Suggestions for the extraction of RNA from cells and tissues are provided, as well as instructions for the generation of labeled targets. Target labeling is achieved by using the sample RNA as a template for the synthesis of cDNA and then generating labeled cRNA in the presence of biotinylated nucleotides. The labeled targets are then spiked with control transcripts to monitor the quality of the subsequent hybridization. Recommendations for washing, staining, and scanning of the arrays are provided.

The steps involved in performing data analysis and verifying data quality measurements are described. The basics of single-array analysis is presented first. This section describes how to obtain qualitative indicators for transcript detection, as well as quantitative measurements of relative abundance. Recommendations for conducting comparative analyses between arrays and new tools for comparing and sharing data are also discussed. Although the application of advanced data analysis techniques depends on the specific goals of individual users, we briefly mention some of the most commonly used approaches.

Experimental design strategies are not discussed in this chapter. However, before starting any microarray project it is important to have a well-defined experiment that is formulated to answer a specific question. The data analysis strategy should also be considered early on during the experimental planning. This will help visualize a clear path to getting and summarizing experimental results. For more information please refer to the Experimental Design, Statistical Analysis, and Biological Interpretation document accessible through the website.

2. Materials

2.1. Equipment

1. Affymetrix scanner system with workstation (Affymetrix; Santa Clara, CA).
2. Fluidics Station (Affymetrix; Santa Clara, CA).
3. Hybridization Oven 640 (Affymetrix; Santa Clara, CA).
4. GeneChip probe array cartridge carriers (Affymetrix; Santa Clara, CA).

2.2. Total RNA Isolation

1. TRIzol Reagent (Invitrogen Life Technologies; Carlsbad, CA).
2. RNeasy Mini Kit (QIAGEN; Valencia, CA).

2.3. cDNA Synthesis

1. SuperScript II (Invitrogen Life Technologies; Carlsbad, CA) or SuperScript Choice System for cDNA Synthesis (Invitrogen Life Technologies; Carlsbad, CA).
2. GeneChip T7-oligo (dT) promoter primer kit.
3. GeneChip Eukaryotic polyA RNA control kit.
4. DEPC-treated water (Ambion, Austin, TX).
5. 5X First Strand cDNA buffer.
6. 0.1 M DTT (Invitrogen Life Sciences, Carlsbad, CA).
7. 10 mM dNTP (Invitrogen Life Technologies; Carlsbad, CA).
8. *E. coli* DNA Ligase (Invitrogen Life Technologies; Carlsbad, CA).
9. *E. coli* DNA Polymerase I (Invitrogen Life Technologies; Carlsbad, CA).
10. *E. coli* RNaseH (Invitrogen Life Technologies; Carlsbad, CA).
11. T4 DNA Polymerase (Invitrogen Life Technologies; Carlsbad, CA).
12. 5X Second strand buffer (Invitrogen Life Technologies; Carlsbad, CA).

2.4. cDNA Cleanup

1. GeneChip Sample Cleanup Module (Affymetrix; Santa Clara, CA).

2.5. Biotin-Labeled cRNA Synthesis

1. GeneChip cRNA labeling kit.

2.6. cRNA Cleanup and Quantitation

1. GeneChip Sample Cleanup Module (Affymetrix; Santa Clara, CA)

2.7. cRNA Fragmentation

1. GeneChip Sample Cleanup Module.

Table 1
Preparation of Hybridization Cocktail for a Single Probe Array

Hybridization Cocktail Components	Final Concentration
Fragmented cRNA	0.05 µg/µL
Control oligonucleotide B2 (3 nM)	50 pM
20X Eukaryotic hybridization controls (*bioB, bioC, bioD, cre*)	1.5, 5, 25 and 100 pM
Herring sperm DNA (10 mg/mL)	0.1 mg/mL
Acetylated BSA (50 mg/mL)	0.5 mg/mL
2X Hybridization buffer	1X

2.8. Hybridization Cocktail

1. Acetylated bovine serum albumin (BSA) solution (50 mg/mL) (Invitrogen Life Technologies; Carlsbad, CA).
2. Herring sperm DNA (Promega Corporation; Madison, WI).
3. GeneChip Eukaryotic Hybridization Control Kit (Affymetrix; Santa Clara, CA).
4. MES Free Acid Monohydrate SigmaUltra (Sigma-Aldrich; St. Louis, MO).
5. MES sodium salt (Sigma-Aldrich; St. Louis, MO).
6. 10% Surfact-Amps 20 (Tween-20), (Pierce Chemical; Rockford, IL).
7. 5 M NaCl, RNAse-free, DNase-free (Ambion, Austin, TX)
8. EDTA Disodium Salt, 0.5 M solution (Sigma-Aldrich; St. Louis, MO).
9. 12X MES stock;1.22 M MES, 0.89 M [Na⁺] (*see* **Note 1**).
10. 2X hybridization buffer ;100 mM MES, 1 M [Na⁺], 20 mM EDTA, 0.01% Tween-20 (*see* **Note 2**).

2.9. Probe Array Washing and Staining

1. *R*-Phycoerythrin streptavidin (Molecular Probes; Eugene, OR).
2. PBS, pH 7.2 (Invitrogen Life Technologies; Carlsbad, CA).
3. 20X SSPE: 3 M NaCl, 0.2 M NaH$_2$PO$_4$, 0.02 M EDTA (Cambrex, East Rutherford, NJ).
4. Goat IgG, reagent grade (Sigma-Aldrich; St. Louis, MO).
5. Biotinylated anti-streptavidin antibody (goat) (Vector Laboratories; Burlingame, CA).
6. Stringent wash buffer; 100 mM MES, 0.1 M [Na⁺], 0.01% Tween-20 (*see* **Note 3**).
7. Non-stringent wash buffer; 6X SSPE, 0.01% Tween-20 (*see* **Note 4**).
8. 2X stain buffer; 100 mM MES, 1 M [Na⁺], 0.05% Tween-20 (*see* **Note 5**).
9. 10 mg/mL goat IgG stock (*see* **Note 6**).
10. The staining and antibody solutions (*see* **Tables 2 and 3**).

3. Methods

The methods described outline the procedure for generating biotinylated cRNA target for expression analysis on eukaryotic GeneChip probe arrays.

Table 2
Preparation of the Staining Solution

SAPE Stain Solution	Final Concentration
2X MES Stain Buffer	1X
50 mg/mL acetylated BSA	2 mg/mL
1 mg/mL Streptavidin-Phycoerythrin	10 µg/mL

Table 3
Preparation of the Antibody Solution

Antibody Solution	Final Concentration
2X MES Stain Buffer	1X
50 mg/mL acetylated BSA	2 mg/mL
10 mg/mL Normal Goat IgG	0.1 mg/mL
0.5 mg/mL biotinylated antibody	3 µg/mL

Please note that these protocols should only be used for eukaryotic organisms owing to the intrinsic differences between eukaryotic and prokaryotic RNA. Prokaryotic-specific guidelines are available through the website, www.affymetrix. com.

A schematic of the gene expression assay, from starting material to probe array scanning, is illustrated in **Fig. 1**.

3.1. Sample Preparation

These protocols are for preparing labeled biotinylated cRNA from total RNA; however, poly (A)$^+$ RNA may be used as starting material with slight modifications.

The first step in the eukaryotic gene expression assay is the purification of RNA from cells or tissues. High-quality starting material is the most crucial component of a successful sample preparation. Therefore, it is important to choose an RNA extraction method that provides the highest quality RNA for the specific tissues or cells being used.

The second step in the protocol is the generation of double-stranded cDNA. Promoter primer T7-(dT) is used in this reaction. This primer facilitates the synthesis of the cDNA strand and incorporates a promoter sequence for use in the the third step of the assay - the in vitro transcription (IVT). After the IVT is complete, the biotin-labeled cRNA is fragmented. This cRNA fragmented target is used to create a hybridization cocktail. The cocktail is hybridized to a

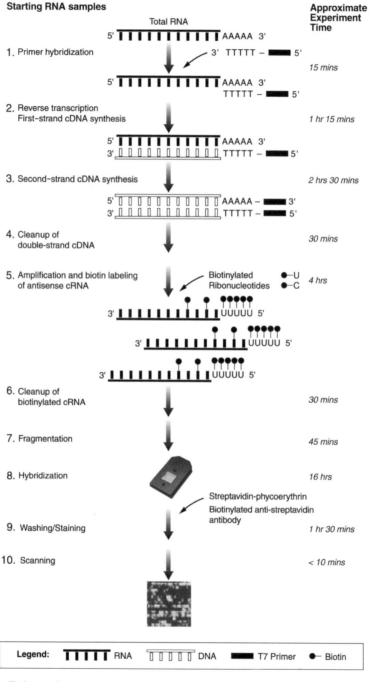

Fig. 1. Eukaryotic gene expression assay, starting from total RNA to the generation of the scanned image (GeneChip Expression Analysis Technical Manual).

GeneChip probe array for 16 h. Next, the array is washed, stained with a fluorescent tag, and scanned using a laser to excite the fluorescent stain. Finally, the captured array image is analyzed using GeneChip software.

3.1.1. Isolation and Quantification of Total RNA

Total RNA isolation from mammalian cells or tissues, Arabidopsis, yeast, and other species can be performed using a variety of methods. As summarized above, it is best to investigate an isolation procedure that is most successful for a particular sample type. RNeasy Total RNA Isolation kit or the TRIzol Reagent provides a robust way for isolation of mammalian and Arabidopsis samples (*see* **Note 7**). When extracting from yeast samples, a hot phenol extraction protocol *(11)* should be considered.

If the RNeasy Total RNA Isolation kit is used, ethanol precipitation is not required, unless concentration of the RNA is necessary. This precipitation is only required when using TRIzol isolation or hot phenol extraction methods.

Prior to proceeding to the cDNA synthesis step, it is important to determine sample concentration and purity by spectrophotometric analysis and gel electrophoresis. The A_{260}/A_{280} ratio should be close to 2.0 for pure RNA, however, ratios between 1.8 and 2.1 are acceptable. RNA degradation is identified by running an agarose gel and examining the 28S and 18S ribosomal RNA (rRNA) bands. These rRNA bands should be clear and with minimal smearing, especially below the 18S band *(12)*. If the RNA purity is not at an acceptable absorbance reading and/or the gel shows signs of smearing, an additional isolation procedure on the RNA samples should be performed. If this does not lead to acceptable quality, then fresh starting material from tissues or cells is required.

The minimum amount of total RNA required for the assay is 5 μg (*see* **Note 8**).

3.1.2. Synthesis of Double-Stranded cDNA From Total RNA

The Invitrogen Life Technologies SuperScript Choice system is required for this section of the assay. However, there are slight modifications to the SuperScript Choice system recommended protocol. For example, a T7-(dT)$_{24}$ oligo primes the first-strand cDNA synthesis in place of oligo (dT) or random primers (*see* **Note 9**).

The recommended amount of starting total RNA for the cDNA protocol is between 5 and 20 μg which subsequently influences the amount of SuperScript II Reverse Transcriptase (200 U/μL) needed. More specifically, if the total RNA starting amount is between 5 and 8 μg, then 1 μL of enzyme is used. If the starting amount of total RNA is between 8.1 and 16 μg , then 2 μL of enzyme is used. Finally, if the starting amount of total RNA is between 16.1 and 20 μg, then 3 μL of enzyme is used.

The first-strand cDNA synthesis involves three steps:

1. Combine the T7-(dT)$_{24}$ primer (final amount 100 pmol), DEPC-H$_2$O and RNA (5–20 µg) mixture and incubate at 70°C for 10 min, spin and place on ice.
2. Add the 5X first strand cDNA buffer (final concentration 1X), 0.1 M DTT (final concentration 10 mM) and 10 mM dNTP mix (final concentration 500 µM each) to the tube and incubate for 2 min at 42°C (*see* **Note 10**).
3. Add the SuperScript II RT enzyme (final content 200–1000 U) to the tube, making the final reaction volume 20 µL. Allow the reaction to proceed for 1 h at 42°C.

When the first-strand reaction is complete, the tube is placed on ice and the second-strand reaction components are added in the following sequence:

1. Add DEPC-H$_2$O and 5X Second-Strand Reaction Buffer (final concentration 1X), 10 mM dNTP mix (final concentration 200 µM each), 10 U/µL *E. coli* DNA Ligase (final content 10 U), 10 U/µL *E. coli* DNA Polymerase I (final content 40 U), 2 U/µL *E. coli* RNase H (final content 2 U). The final volume, first strand plus second strand, should be 150 µL.
2. Gently tap the tube to mix and briefly microcentrifuge to remove any condensation. Then, incubate at 16°C for 2 h in a cooling water bath.

After the second-strand synthesis is complete, add 2 µL of T4 DNA Polymerase (10 U) and return tube to 16°C for 5 mins. Then, add 10 µL 0.5 M EDTA to stop the reaction.

The reaction can be stored at −20°C for later use (*see* **Note 11**).

3.1.3. Cleanup of Double-Stranded cDNA

The cleanup of the double-stranded cDNA reaction is imperative to rid the sample of impurities. This step is accomplished by using Phase Lock Gels or a column purification method such as the GeneChip Sample Cleanup Module. If using the Phase Lock gels, be sure to ethanol precipitate the samples after purification before going to the next step. Ethanol precipitation is not required when using the column purification method.

3.1.4. Synthesis of Biotin-Labeled cRNA

The Enzo BioArray HighYield (HY) RNA Transcript Labeling Kit is used to generate biotin-labeled cRNA. This reaction is catalyzed by the addition of T7 RNA Polymerase, which recognizes the promoter region incorporated into the sequence during the first-strand cDNA synthesis reaction. This IVT reaction generates a 50- to 100-fold linear amplification of the represented transcripts (*see* **Note 12**).

The amount of cDNA used in the IVT reaction depends on the original amount of starting material. More specifically, if the starting total RNA isolated is between 5.0 and 8.0 µg, 10 µL of cDNA should be used. If the starting total RNA is between 8.1 and 16.0 μg of total RNA, 5 μL should be used. If the starting total RNA is between 16.1 and 20 μg, 3.3 μL of cDNA should be used. The reaction components are added to the cDNA target along with the appropriate amount of water. The final reaction volume is 40 μL (*see* **Note 13**).

Once the reagents are added, the tube is mixed gently, microcentrifuged briefly for 5 s, and quickly placed in a 37°C water bath for 4–5 h. Mix the reaction every 30–45 min during the incubation. The labeled cRNA can be stored at −20° or at −70°C for long-term storage (*see* **Note 14**).

3.1.5. In Vitro Transcription Cleanup

Cleaning the products of the IVT rids the sample of excessive nucleotides, salts, and other impurities. Accomplish this step by using the GeneChip Sample Cleanup Module.

3.1.6. cRNA Quantification

It is imperative to determine the purity and yield of the cRNA target through spectrophotometric analysis and gel electrophoresis. Acceptable A_{260}/A_{280} absorbance ratios are between 1.8 and 2.1. If a sample does not meet this criterion, it is advisable to repeat the experiment. Gel electrophoresis provides an illustration of the yield and size distribution of the labeled target.

Another step in quantifying the cRNA yield is to account for unlabeled RNA in the reaction. Unlabeled RNA is accounted for by adjusting the cRNA yield using the following equation:

Adjusted cRNA yield (μg) = (cRNA yield after IVT) − (RNA starting amount) * (cDNA used in the IVT)

3.1.7. cRNA Fragmentation

The cRNA is fragmented by a metal-induced hydrolysis process which segments the target into fragments ranging from 35 to 200 bases. It is important to have the correct concentration of the reaction components - cRNA, fragmentation buffer, and water, as well as ensuring that the time and temperature are exactly those recommended. The maximum amount of cRNA to fragment depends on the volume of the hybridization cocktail, which ultimately depends on the size of the array. For example, for a standard array, the minimum amount to fragment is 10 μg of cRNA for a 200 μL cocktail.

Fragmentation buffer (5X), cRNA, and water is added to the reaction to make a total volume of 40 μL (*see* **Note 15**). The reaction is incubated at 94°C for 35 min. The tube is then placed on ice or stored at −20°C until the hybridization procedure. An aliquot of fragmented cRNA is saved for gel analysis, so that the fragmented target can be compared to the purified and unpurified cRNA.

3.2. Sample Hybridization
and Probe Array Washing, Staining, and Scanning

3.2.1. Hybridization Cocktail

The hybridization cocktail includes the fragmented cRNA target, 20X Eukaryotic Hybridization Controls (*E.coli bioB*, *bioC*, *bioD* and bacteriophage *cre* controls), Oligo B2, acetylated BSA, and Herring Sperm DNA (*see* **Note 16**).

Mix the following reagents with buffered solution for a final volume that varies depending on the array type and the number of hybridizations. Be sure to heat the 20X Eukaryotic Controls at 65°C for 5 min in order to resuspend the mixture completely.

Once the hybridization cocktail is prepared, the probe arrays are taken out of 4°C and equilibrated to room temperature. At the same time, the hybridization cocktail is heated to 99°C for 5 mins and then transferred to another 45°C heat block for 5 mins. The cocktails are then spun at maximum speed in a microcentrifuge for 5 mins to separate any insoluble material from the qualified hybridization mixture. Meanwhile, the arrays are prehybridized with 1X hybridization buffer. The buffer is injected into the lower septa of the array and the upper septum is vented for air release. The probe arrays are then incubated in the hybridization oven for 10 mins at 45°C at a rotation speed of 60 rpm. Once prehybridization is complete, the buffer solution is removed from the probe array cartridge and the array is filled with approx 80% of the hybridization cocktail solution (*see* **Note 17**). The probe arrays are balanced and placed in the hybridization oven for 16 h at 45°C.

3.2.2. Preparation for Probe Array Washing and Staining

After the 16-h hybridization, the cocktail is removed from the probe array and saved. The cocktail can be stored at −20°C or at −80°C (*see* **Note 18**).

Once the sample is removed, the probe array is filled completely with nonstringent wash buffer.

The following steps prepare the array for an automatic washing and staining procedure performed on the GeneChip Fluidics Station:

1. Open the GeneChip System Workstation.
2. Turn on the fluidics machine and scanner.
3. Create an experiment file (.EXP) in GeneChip software for each probe array.
4. Prime the fluidics machine with the appropriate wash buffers (nonstringent and stringent).
5. Prepare the streptavidin-phycoerythrin (SAPE) staining and antibody solutions (*see* **Note 19**).

The staining procedure used for most GeneChip probe arrays requires a staining and an antibody amplification step. This process starts by staining the array with SAPE, which recognizes the biotin-labeled ribonucleotides. A second solution, which includes an anti-streptavidin biotinylated antibody, is washed over the array. Finally, another solution of SAPE is added to the array that binds to the biotinylated antibody and provides further amplification of the signal.

Add deionized water to the SAPE stain solution for a final volume of 600 µL. This reaction can be doubled, in order to make a master mix that is enough for both of the SAPE stains.

Add deionized water to the antibody solution for a final volume of 600 µL.

3.2.3. Fluidics Washing and Staining

The probe array is washed and stained on the fluidics machine using array-specific protocols recommended by Affymetrix. For example, the fluidics protocol EukGE-WS2 is used for the standard format array. The name of the protocol indicates that the array is for eukaryotic (Euk) gene expression (GE) samples that go through two washing and staining (WS2) procedures. The protocol takes approx 75 mins to complete. The majority of the fluidics protocols consist of the following steps:

1. 10 cycles of 2 mixes per cycle with nonstringent buffer (*see* **Subheading 2.**) at 25°C.
2. 4 cycles of 15 mixes per cycle with stringent buffer (*see* **Subheading 2.**) at 50°C.
3. SAPE stain for 10 mins at 25°C.
4. 10 cycles of 4 mixes per cycle with nonstringent buffer at 25°C.
5. Antibody stain for 10 mins at 25°C.
6. SAPE stain for 10 mins at 25°C.
7. 15 cycles of 4 mixes per cycle with nonstringent buffer at 30°C.

Once the fluidics protocols are complete, check the probe array for bubbles. Bubbles occur when the nonstringent buffer does not completely fill the probe array chamber during the final fill step. If bubbles are present, return the array to the probe array holder to automatically perform a drain and fill. If this does not remove the bubbles, this step needs to be performed manually by pipeting nonstringent buffer into the array chamber. Ensure that all bubbles are removed before scanning and that the glass surface is clean and free of dust, lint, and other materials that can interfere with the scanning procedure. If the glass needs to be cleaned, use a non-abrasive towel or tissue to gently wipe the glass surface before scanning.

Once the fluidics protocol is completed and each array is checked for bubbles, the fluidics machine is cleared of buffer and other contaminants by performing a shutdown procedure.

3.2.4. Scanning

The GeneChip scanner must be turned on 15 mins prior to use. The scan time takes approx 10 mins depending on the array type. The scanned data is represented

Lescallett et al.

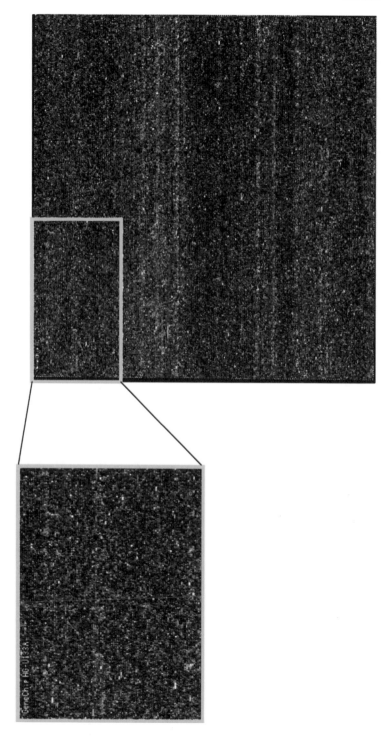

Fig. 2. Screen shot of the microarray scanned image representing the intensity value for each probe cell.

as a .DAT or image file and saved on the computer (**Fig. 2**). Immediately following the creation of a .DAT, the software automatically creates a .CEL file. This file contains a single intensity value for each probe cell.

3.3. Data Analysis

3.3.1. Single-Array Data Analysis

Whether classifying samples based on their expression profiles, identifying transcripts of potential biological or medical importance, or building expression databases, most array experiments involve working with data obtained from multiple arrays. The consistency and reproducibility of GeneChip arrays uniquely positions this platform to achieve these comparisons. Before integrating these data sets, however, the results generated by single arrays must be reviewed and processed. This section describes a basic procedure for analyzing data from single arrays, applicable to many experimental situations. Depending on specific experimental techniques and goals; however, users may need to modify these guidelines.

Open the Affymetrix software and view the scanned image(s) (.dat file). Check for image artifacts such as high or low density spots, uneven background, or other abnormalities. Apply a grid and enlarge each of the four corners of the array image to check the intensity and grid alignment of the control Oligo B2 hybridization (*see* **Note 20**). Next, adjust the expression analysis settings so that scaling, normalization, probe mask, baseline, and the algorithm defaults are set appropriately. If experimental samples are going to be compared to a baseline or control sample, it is important to choose a scaling or normalization method that best fits the experimental design. For example, if the majority of transcripts in an experimental sample are not expected to change compared to a control, then a global scaling approach is a suitable strategy. Conversely, when a large number of changes are expected to occur between the experiment and control samples, an approach that scales to a selected number of uniformly expressed transcripts is recommended (*see* **Note 21**).

In both global and selected scaling methods, an arbitrary number, called "target intensity," is used across all experiments, allowing interexperiment comparisons. This number facilitates the generation of a scaling factor by which each signal value on the array is multiplied.

3.3.2. The Detection Algorithm

After these preparation steps, the data analysis output or .CHP file is generated. This file contains detection calls, indicators of whether a transcript is reliably detected or not, and signal values, relative measures of transcript abundance. The following section briefly explains how these outputs are generated.

Transcript or probe set detection and quantification depends on analyzing the hybridization signals of the 11–20 probe pairs. These probe pairs represent different 25-mer segments of a particular transcript. For each segment or probe that is perfectly identical to a target sequence, GeneChip arrays provide a partner probe that is identical except for a single base mismatch at the 13th position. These probe pairs, containing the perfect match (PM) and mismatch probes (MM), allow for the assessment of real and stray (nonspecific) signals across the probe set.

The detection algorithm uses a nonparametric test, based on a one-sided Wilcoxon signed rank, to evaluate probe pair intensities and generate a detection *p*-value with an associated present (P), marginal (M), or absent (A) call (*see* **Note 22**). The first step in determining the *p*-value is calculating the discrimination score (R). The discrimination score is an indicator of target-specific intensity differences between the perfect match and the mismatch and is calculated as:

$$R = (PM - MM) / (PM + MM)$$

Each probe pair discrimination score is then adjusted by an empirically derived, small positive number called Tau (*see* **Note 23**). The adjusted discrimination scores are then ranked according to the absolute value. Once ranked, the sign is re-applied, the positive rank values are summed, and a *p*-value is generated. Individual transcripts are assigned a P, M or A call based on user-defined, *p*-value cut-offs known as $\alpha 1$ and $\alpha 2$ (*see* **Note 24**). Values falling below $\alpha 1$ are assigned a P call, those between $\alpha 1$ and $\alpha 2$ an M call, and those above $\alpha 2$ an A call. The final output is a call with an associated *p*-value.

3.3.3. The Signal Algorithm

The relative level of expression for each transcript is calculated using an algorithm based on the one-step Tukeys biweight estimate. This robust method provides an effective approach to handling outliers that, instead of being dropped, are smoothly down weighted.

The first step in the process of deriving signal is to identify the median of the data. This is done by calculating the log of the PM intensity after subtracting the stray signal estimate, obtained from the MM intensity or the idealized MM intensity (*see* **Note 25**). The closer this value is to the median value of the set, the more strongly it is weighted. The mean is then calculated once all of the pairs have been weighted. The weighted mean is converted back to the linear scale and the output is a quantitative metric called signal.

3.3.4. Quality Control

Generating an expression analysis report file (.RPT) derived from the analysis output file (.CHP) can perform most of the quality review of an array experiment. The report allows users to assess sample quality, assay execution, and hybridization performance. The results from the control *bioB* transcripts, included in the hybridization cocktail at 1.5 [p*M*], offer an indication of the assay's sensitivity. In a typical experiment, *bioB* should be called P most of the time. *BioC*, *bioD*, and *cre* should always be called P and should show increasing signal values that correspond to their relative concentrations.

RNA sample and assay quality are often monitored by comparing the signal values of the 3' probe sets to the 5' probe sets of actin and GAPDH transcripts. Given that the assay for generating labeled targets has an intrinsic 3' bias, because of the reverse transcription from the 3' polyA tail, the ratio of 3' to 5' signal values is usually greater than 1. However, ratios that exceed three indicate either degraded sample RNA or inefficient IVT (*see* **Note 26**).

Another indicator of sample quality can be the percentage of probe sets assigned a P call. This percentage varies depending on biological factors, such as cell or tissue type, but extremely low values may indicate poor sample quality. The percentage is also useful for assessing the reproducibility of replicate experiments.

The average background and raw noise values should also be inspected. Although background can vary widely, average background values typically fall between 20 and 100. Ideally, arrays should have similar background levels if they are being compared. The noise value, a measure of pixel-to-pixel variation, should also be similar. Although sample quality can contribute to noise, usually the most significant contributor is the electrical noise from the scanner.

It is important to keep a running log of the quality control metrics for each sample in order to monitor sample performance and identify sample outliers.

3.3.5. Viewing the Data

After reviewing the report file, return to the .CHP file. The signal values, detection calls, and detection *p*-values for each transcript can be viewed and sorted according to user preferences (**Fig. 3**). The data can also be imported as a text file into other programs, such as Microsoft® Excel™.

3.4. Array Comparison Analysis

The goal of many gene expression experiments is to compare the transcription profiles of two samples. To begin analysis, obtain a .CHP file for each of the samples to be compared. Designate one of the arrays as the baseline, and the other as the experimental array (the choice can be arbitrary, but should be used consistently throughout subsequent analyses) (*see* **Note 27**). The difference values (PM-MM) of each probe pair in the baseline array are compared to their matching probe pairs in the experimental array. As in single-array analysis, comparison analysis involves two algorithms that generate a qualitative

	Stat Pairs	Stat Pairs Used	Signal	Detection	Detection p-value
37984_s_at	16	16	92.2	P	0.000218
32102_at	16	16	59.5	P	0.000218
37900_at	16	16	72.6	P	0.000219
31697_s_at	16	16	664.2	P	0.000219
40567_at	16	16	502.3	P	0.000219
35808_at	16	16	212.6	P	0.000219
34819_at	16	16	143.0	P	0.000219
35787_at	16	16	295.7	P	0.000219
35758_at	16	16	301.0	P	0.000219
34817_s_at	16	16	339.6	P	0.000219
34644_at	16	16	723.9	P	0.000219
34608_at	16	16	3313.0	P	0.000219

Fig. 3. Data analysis output (.CHP file) for a Single-Array Analysis includes Stat Pairs, Stat Pairs Used, Signal, Detection, and Detection *p*-value for each probe set.

output with an associated *p*-value, and a quantitative metric, also associated with a confidence interval (CI). The qualitative output is called the change call, which indicates if a transcript in the experimental array is increased, decreased, or equivalent to its baseline counterpart. The quantitative metric is called the signal log ratio and is a quantitative estimate of the change in gene expression.

3.4.1. Change Algorithm

Similar to single-array analyses, comparison analyses rely on a Wilcoxon rank test. First, each probe pair is evaluated for intensity saturation. Then, each probe set in the experimental array is compared to the matching set in the baseline array to generate a change *p*-value. User-defined cut-off values, called gammas, are then applied to the *p*-values to generate discrete change calls (increase [I], marginal increase [MI], no change [NC], marginal decrease [MD], or decrease [D]). P-values range from 0.0 to 1.0, with those close to 0.0 indicating a probable increase in the experimental probe set relative to the baseline set, and those close to 1.0 indicating a likely decrease. Values close to 0.5 indicate probe sets whose intensities are very similar in the baseline and experimental data sets.

3.4.2. Signal Log Ratio Algorithm

The Signal Log Ratio provides an estimate of the magnitude and direction of change in transcript abundance between two arrays. Like the signal value

	Stat Common Pairs	Signal Log Ratio	Signal Log Ratio Low	Signal Log Ratio High	Change	Change p-value
35839_at	16	0.3	0.2	0.4	I	0.000014
1799_at	16	0.9	0.5	1.3	I	0.000015
35985_at	16	0.4	0.3	0.5	I	0.000015
34696_at	16	0.4	-0.1	0.9	I	0.000023
31356_at	16	1.8	0.8	2.8	I	0.000025
35202_at	16	0.4	0.2	0.6	I	0.000027
39651_at	16	0.4	0.3	0.5	I	0.000029
39777_at	16	0.4	0.1	0.6	I	0.000031
37610_at	16	0.4	0.2	0.5	I	0.000034
32070_at	16	0.3	0.2	0.4	I	0.000034
1581_s_at	16	0.7	0.1	1.3	I	0.000037
35283_at	16	0.5	0.3	0.6	I	0.000037

Fig. 4. Data analysis output (.CHP file) for a Comparison Analysis includes Stat Common Pairs, Signal Log Ratio, Signal Log Ratio Low, Signal Log Ratio High, Change, and Change *p*-value for each probe set.

derived from single-array analyses, the log ratio is calculated using a one-step Tukeys biweight method. The log ratio algorithm calculates a mean of the log ratios of probe pair intensities across two arrays (*see* **Note 28**). Ninety-five-percent CIs are also calculated to provide a measure of the variation in the biweight estimate. Small CI indicate that the data are less variable and more accurate.

3.4.3. Viewing the Data

After reviewing the report file, return to the .CHP file. The signal log ratio, change calls, and change *p*-values for each transcript on the experimental sample can be viewed and sorted according to user preferences (**Fig. 4**). The data can also be imported as a text file into other programs, such as Microsoft Excel.

3.5. Advanced Data Analysis and Mining

It is beyond the scope of this chapter to provide an in-depth guide to advanced microarray data analyses, but this section offers some general pointers regarding the available tools. A variety of algorithms have been described to group samples or genes with similar expression patterns. Clustering analyses are often used in studies aimed at discovering new disease classes or novel relationships between genes. These methods rely on unsupervised algorithms, which search for patterns of gene expression without taking into account any previously known biological, clinical, or demographic information. Although some of these algorithms allow users to impose a few constraints on the clusters generated *(13)*, the main advantage of clustering is the ability to provide systematic and unbiased analyses of expression data. Studies using self-organizing maps (SOMs) *(13)*, hierarchical algorithms *(14)*, and k-means clustering algorithms *(15)* illustrate the capabilities of such techniques.

For some applications, however, supervised algorithms that incorporate prior knowledge into the analyses are more useful. These algorithms can be "trained" to search for expression patterns associated with particular traits, such as disease outcomes or responsiveness to drugs, and then used to predict those traits in new, unknown samples. Examples include k-nearest neighbors algorithms *(5)*, weighted voting algorithms *(16,17)*, the support vector machine method *(18)*, Bayesian models *(19)*, and artificial neural networks *(20)*.

Whether applying supervised or unsupervised algorithms, however, users should be aware of the problem of "multiple comparisons." Given the large number of results per array experiment, even a small percentage of false positives can result in a large absolute number of artifactual correlations. To minimize this problem, many investigators set aside samples for conducting independent tests, and apply permutation tests in which they introduce noise or scramble the data and then assess how much the identified correlations differ from correlations that could arise by chance. Although these statistical tests are powerful, it is important to note that expression patterns may still result from random associations.

3.6. Data Management

The number of genes that can be simultaneously monitored with the GeneChip platform is unequalled. Because GeneChip arrays generate large amounts of data it is critical to set up consistent procedures for data storage and handling. Deciding on a clear and concise nomenclature for each project, performing regular back-ups of all files, and employing database management software are highly recommended.

Affymetrix has developed software that employs a centralized data management system for moderate to high throughput laboratories. This software facilitates data sharing among groups, allows automation of data analysis, has more sophisticated security capabilities, and increases throughput by liberating workstations from analysis tasks.

An important feature of both systems is that they provide the flexibility of open architecture design, allowing users to access a wide variety of tools for analyzing and exchanging data. This flexibility derives from the Affymetrix

Analysis Data Model (AADM), a relational database schema that stores array results in a format that can be easily recognized and used by many software programs. Four related subschema hold the data associated with each experiment: array design (which includes information about the array, such as its numbers of rows and columns), experiment setup (including information about the target applied), analysis results (ranging from individual cell intensities to comparative analysis results), and protocol parameters. AADM's open design is proving particularly useful in light of the growing number of analytical algorithms being developed in academia and industry, and users' increasing need to share and compare their data.

An additional software tool that complements the flexibility of AADM-based databases is NetAffx Analysis Center at Affymetrix.com. Through this online center, array users can efficiently collect and integrate a wide variety of information relevant to their specific experimental results and aims. This site provides access to a variety of public databases, including GenBank, dbEST, RefSeq, and UniGene. In addition, it links users to proprietary databases that offer annotations, such as protein domain alignments, as well as target and probe sequences for GeneChip arrays. Researchers can use the site to search array probe sets for particular sequences, review gene and protein annotations, and sort transcripts by a number of criteria, such as functional groups, metabolic pathways, or disease association. The Gene Ontology Mining Tool provides visualization mapping of probe sets to gene groups in detail or at a broad level.

4. An Array of Possibilities

A wealth of studies illustrate how the guidelines described in this chapter can be used to answer a variety of biological and medical questions. Applications range from probing biological processes, such as development *(21,22)* and circadian rhythms *(23,24)*, to searching for predictors of disease and drug responsiveness *(25)*. Cancer research is a rapidly growing field of application, in which arrays have helped investigators discover new tumor classes, assign patient samples to known tumor classes, predict clinical outcomes, reveal cancer-associated alterations in molecular pathways, and identify new drug targets *(26)*.

In one of the most comprehensive leukemia studies to date, for example, Yeoh and co-workers used GeneChip Human Genome U95A arrays to monitor the expression of more than 12,600 genes in leukemic blasts from 360 pediatric ALL patients *(6)*. The study showed that through expression profiling, it is possible to not only classify all known leukemia subtypes that are prognostically relevant, but to identify patients that are at risk of failing conventional treatments. In addition, the array data supplied molecular candidates for developing new treatments, as well as suggested new diagnostic and subclassification markers. As often occurs when applying microarray techniques, the authors were able to extract valuable information about the whole genome relevant to multiple questions from their data sets.

5. Notes

1. 1000mL 12X MES Buffer
   70.4 g MES free acid monohydrate
   193.3 g MES Sodium Salt
   800 mL of Molecular Biology Grade water
   Mix and adjust volume to 1000 mL
   The pH should be between 6.5 and 6.7; pass through a 0.2 μm filter.3.
2. 50 mL 2X Hybridization Buffer
   8.3 mL of 12X MES Stock
   17.7 mL of 5 *M* NaCl
   4.0 mL of 0.5 *M* EDTA
   0.1 mL of 10% Tween-20
   19.9 mL of water
   Store at 2–8°C, and shield from light
3. 1000 mL Stringent wash buffer
   83.3 mL of 12X MES stock buffer
   5.2 mL of 5 *M* NaCl
   1.0 mL of 10% Tween-20
   910.5 mL of water
   Pass through a 0.2 μm filter
   Store at 2–8°C and shield from light
4. 1000 mL Nonstringent wash buffer
   300 mL of 20X SSPE
   1.0 mL of 10% Tween-20
   699 mL of water
   Pass through a 0.2 μm filter
5. 250 mL 2X Stain buffer
   41.7 mL 12X MES Stock buffer
   92.5 mL 5 *M* NaCl
   2.5 mL 10% Tween-20
   113.3 mL water
   Pass through a 0.2 μm filter
   Store at 2–8°C and shield from light
6. 10 mg/mL Goat IgG Stock
   Resuspend 10 mg in 1 mL 150 m*M* NaCl
   Store at 4°C
7. When TRIzol is used to isolate total RNA it is recommended that a second cleanup on the total RNA is performed in order to obtain sufficient cRNA yields. This can be done with QIAGEN RNeasy Total RNA isolation kit.

8. The required amount of poly(A)$^+$ starting material is 0.2–2.0 µg. There is a small sample protocol that can be used for limiting amount of starting total RNA material, please refer to www.affymetrix.com or to the GeneChip Expression Analysis Technical Manual.

9. The oligo T7-(dT)$_{24}$ primer (5' GGCCAGTGAATTGTAATACGACTCACTATAGGGAGGCGG-(dT)$_{24}$-3', 100 pmol/µL) must be HPLC purified to achieve efficient cDNA synthesis and in vitro transcription. Poorly made primer will lead to lower cRNA yield.

10. If Poly (A)$^+$ is used, it is important to adjust the temperature of the first-strand cDNA synthesis to 37°C from 42°C used for total RNA.

11. RNase treatment of the cDNA prior to the in vitro transcription is not recommended.

12. Prior to use, centrifuge all reagents briefly to ensure that the components remain at the bottom of the tube. The product should not be used after the expiration date stated in the label. If precipitation occurs in the reaction buffer, centrifuge briefly to remove precipitate before use. The precipitation does not interfere with the reaction.

13. The amount of cDNA used in the in vitro transcription reaction for poly (A)$^+$ RNA varies from the amount of total RNA used.

14. It is useful to save an aliquot of the unpurified IVT reaction for analysis by gel electrophoresis.

15. The cRNA in the fragmentation reaction must be at a final concentration range of 0.5–2.0 µg/µL. If the sample is more dilute, perform an ethanol precipitation step before proceeding.

16. When preparing the hybridization cocktail, it is important to consider the probe array type being used because different arrays require different amounts of cRNA.

17. While pipeting the solution, be sure to avoid any insoluble material at the bottom of the tube.

18. Once the hybridization cocktail is pipeted out of the array and the array chamber is filled with the nonstringent buffer, it is possible to store the array at 4°C for up to 4 h before proceeding to the washing and staining steps. Be sure to equilibrate the probe array to room temperature before washing and staining.

19. Always store the SAPE reagent in the dark at 4°C (do not freeze). Be sure to mix the SAPE thoroughly, but gently, before adding to the rest of the reaction components. Always prepare the SAPE stain solution immediately before use.

20. The control oligonucleotide B2 should generate hybridization signals that trace the boundaries of the probe area. The controls appear as an alternating pattern of intensities with a checkerboard pattern at each corner and spell out the name of the array. In addition to serving as a positive control, the pattern is used by the software to align the array image with a grid. If the intensity of the checkerboard patterns is too high or too low, or if the pattern is distorted, the grid must be aligned manually.

21. One option is to apply a normalization method based on the intensities of 100 control probe sets.

22. To establish whether a transcript is present in detectable amounts, evaluate the level of signal saturation for each probe pair. If a MM probe is saturated (46,000 for the 2500 GeneArrayScanner), the signal from the corresponding PM probe is uninformative, and the probe pair is discarded.

23. The default value of Tau is set at 0.015. Tau can be adjusted to balance sensitivity and specificity. If the experiment is designed to achieve high sensitivity and avoid false negatives, while tolerating some miscalls, Tau can be decreased. If the experiment is designed to achieve high specificity, avoiding false positives, while missing a few positive calls, Tau can be increased.

24. $\alpha 1$ and $\alpha 2$ default values change depending on the number of probe pairs.

25. The signal algorithm is designed to avoid generating negative signal values, which lack physiological meaning and can interfere with subsequent data processing. If a MM value is higher than a PM value, as a result of cross-hybridization, the uninformative MM is replaced with either an adjusted MM value calculated from the mean of the PM:MM ratios of the other probes in the set, or a value that is slightly lower than the PM and which results in an absent call.

26. If only one of the controls has a ratio above 3, do not automatically assume that the quality of the experimental data is compromised. The elevated ratio may be the result of transcript specific changes rather than low sample or assay quality. It is important to compare the outcomes of the various quality indicators, as well as accumulation of previous experiment results, before reaching a final assessment.

27. Before running an analysis, check the Expression Analysis Settings with particular attention to the scaling or normalization criteria.

28. Logarithms are used because hybridization behavior is best described by exponential functions. In addition, signal log ratios can provide more sensitive indicators of the differences between probe values than linear -fold changes. When the experimental and baseline values are very similar, log ratios outperform fold-change measurements. In addition, because the log scale used by the algorithm is base 2, the Signal Log Ratio is easily converted to a fold-change value, if desired. A value of 1.0 indicates a twofold increase, a value of -1.0 indicates a twofold decrease, and a value of 0 indicates no change at all. The algorithm also provides an estimate of the amount of variation in the data in the form of CIs, which are calculated based on the variation between probes in a set.

Acknowledgments

Some of the material in this review was derived from the Affymetrix GeneChip Expression Analysis Technical Manual. We are indebted to all who participated in its production. We would also like to thank Brian Shimada, Raji Pillai, Bob Kolovch, and Yan Zhang-Klompus for their editorial suggestions.

References

1. Lockhart, D. J. and Winzeler, E. A. (2000) Genomics, gene expression and DNA arrays. *Nature* **405,** 827–836.

2. Notterman, D. A., Alon, U., Sierk, A. J., and Levine, A. J. (2001) Transcriptional gene expression profiles of colorectal adenoma, adenocarcinoma, and normal tissue examined by oligonucleotide arrays. *Cancer Res.* **61,** 3124–3130.

3. Tice, D. A., Szeto, W., Soloviev, I., et al. (2002) Synergistic induction of tumor antigens by wnt-1 signaling and retinoic acid revealed by gene expression profiling. *J. Biol. Chem.* **277,** 14329–14335.

4. Ferrando, A., Neuberg, D., Staunton, J., et al. (2002) Gene expression signatures define novel oncogenic pathways in T cell acute lymphoblastic leukemia. *Cancer Cell* **1,** 75–87.

5. Pomeroy, S. L., Tamayo, P., Gaasenbeek, M., et al. (2001) Gene expression-based classification of outcome prediction of central nervous system embryonal tumors. *Nature* **415,** 436–441.

6. Yeoh, E. J., Ross, M., Shurtleff, S., et al. (2002) Classification, subtype discovery, and prediction of outcome in pediatric acute lymphoblastic leukemia by gene expression profiling. *Cancer Cell* **1,** 133–143.

7. MacDonald, T. J., Brown, K. M., LaFleur, B., et al. (2001) Expression profiling of medulloblastoma: PDGFRA and the ras/mapk pathway as therapeutic targets for metastatic disease. *Nat. Genet.* **29,** 143–152.

8. Lockhart, D. J., Dong, H., Byrne, M. C., et al. (1996) Expression monitoring by hybridization to high-density oligonucleotide arrays. *Nat. Biotechnol.* **14,** 1675–1680.

9. Gerhold, D., Lu, M., Xu, J., Austin, C., Caskey, C. T., and Rushmore, T. (2001) Monitoring expression of genes involved in drug metabolism and toxicology using DNA microarrays. *Physiol. Genomics* **5,** 161–170.

10. Fodor, S. P. A., Read, J. L., Pirrung, M. C., Stryer, L., Lu, A. T., and Solas, D. (1991) Light-directed, spatially addressable parallel chemical synthesis. *Science* **251,** 767–773.

11. Schmitt, M. E., Brown, T. A., and Trumpower, B. L. (1990) A rapid and simple method for preparation of RNA from *Saccharomyces cerevisiae. Nucleic Acids Res.* **18,** 3091–3092.

12. Farrell, R. (1998) RNA Methodologies, Academic Press.

13. Tamayo, P., Slonim, D., Mesirov, J., et al. (1999) Interpreting patterns of gene expression with self-organizing maps: Methods and application to hematopoietic differentiation. *Proc. Natl. Acad. Sci. USA* **96,** 2907–2912.

14. Eisen, M. B., Spellman, P. T., Brown, P. O., and Botstein, D. (1998) Cluster analysis and display of genome-wide expression patterns. *Proc. Natl. Acad. Sci. USA* **95,** 14863–14868.

15. Tavazoie, S., Hughes, J. D., Campbell, M. J., Cho, R. J., and Church, G. M. (1999) Systematic determination of genetic network architecture. *Nat. Genet.* **22,** 281–285.

16. Golub, T. R., Slonim, D. K., Tamayo, P., et al. (1999) Molecular classification of cancer: class discovery and class prediction by gene expression monitoring. *Science* **286,** 531–537.

17. Shipp, M., Tamayo, P., Ross, K., et al. (2002) Diffuse large B-cell lymphoma outcome prediction by gene expression profiling. *Nat. Med.* **8,** 68–74.

18. Brown, M. P., Grundy, W. N., Lin, D., et al. (2000) Knowledge-based analysis of microarray gene expression data by using support vector machines. *Proc. Natl. Acad. Sci. USA* **97,** 262–267.

19. West, M., Blanchette, C., Dressman, H., et al. (2001) Predicting the clinical status of human breast cancer by using gene expression profiles. *Proc. Natl. Acad. Sci. USA* **98,** 11462–11467.

20. Khan, J., Wei, J. S., Ringner, M., et al. (2001) Classification and diagnostic prediction of cancers using gene expression profiling and artificial neural networks. *Nat. Med.* **7,** 673–679.

21. Muller, H., Bracken, A. P., Vernell, R., et al. (2001) E2Fs regulate the expression of genes involved in differentiation, development, proliferation, and apoptosis. *Genes Dev.* **15,** 267–285.

22. Mody, M., Cao, Y., Cui, Z., et al. (2001) Genome-wide gene expression profiles of the developing mouse hippocampus. *Proc. Natl. Acad. Sci. USA* **98,** 8862–8867.

23. Storch, K. F., Lipan, O., Leykin, I., et al. (2002) Extensive and divergent circadian gene expression in liver and heart. *Nature* **417,** 78–83.

24. Ueda, H. R., Matsumoto, A., Kawamura, M., Iino, M., Tanimura, T., and Hashimoto, S. (2002) Genome-wide transcriptional orchestration of circadian rhythms in *Drosophila. J. Biol. Chem.* **277,** 14048–14052.

25. Chicurel, M. and Dalma-Weiszhausz, D. (2002) Microarrays in pharmacogenomics: Advances and future promise. *Pharmacogenomics* **5,** 589–601.

26. Chicurel, M. E. and Dalma-Weiszhausz, D. D. (2003) Oligonucleotide Microarrays. In: Expression profiling of human tumors (Ladanyi, M. and Gerald, W. L., eds.), Humana Press, Inc., Totowa, NJ.

7

Amplified Differential Gene Expression Microarray

Zhijian J. Chen and Kenneth D. Tew

Summary

Amplified Differential Gene Expression (ADGE) and DNA microarray provides a new concept that the ratios of differentially expressed genes are magnified prior to detecting them. The ratio magnification is achieved with the integration of DNA reassociation and polymerase chain reaction (PCR) amplification and ensured with the design of the adapters and primers. The ADGE technique can be used either as a stand-alone method or in series with DNA microarray. ADGE is used in sample preprocessing and DNA microarray is used as a displaying system in the series combination. The combination of ADGE and DNA microarray provides a mutual complement of their strengths: the magnification of ratios of differential gene expression improves the detection sensitivity; the PCR amplification and efficient labeling enhance the signal intensity and reduce the requirement for large amounts of starting material; and the high throughput for DNA microarray is maintained.

Key Words: ADGE, amplified differential gene expression, DNA microarray, gene expression

1. Introduction

Amplified Differential Gene Expression (ADGE) is designed to quadratically magnify the ratios of genes in two samples prior to detection *(1)*. The schematic procedure is shown in **Fig. 1**. Two comparative nucleic acid samples are selected, one for control, the other for tester. The control and tester cDNA are cut with *Taq*I after they are synthesized from total RNA or mRNA. The *Taq*I fragments of control and tester DNA are ligated to the CT adapter and TT adapter, respectively. The adapter-linked control and tester DNA are reassociated through mixing in equivalent amounts, denaturing, and then annealing. From

From: *Methods in Molecular Biology, Vol. 258: Gene Expression Profiling: Methods and Protocols*
Edited by: R. A. Shimkets © Humana Press Inc., Totowa, NJ

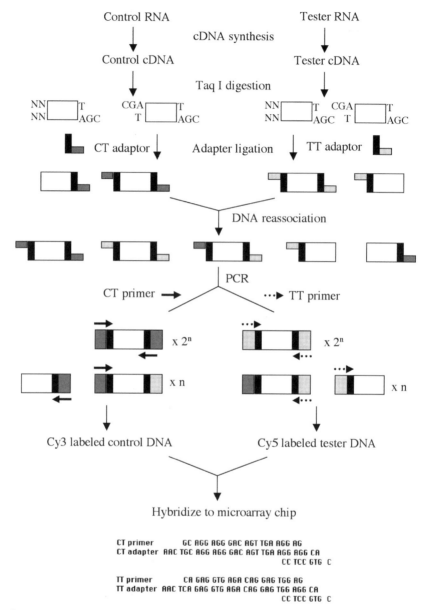

Fig. 1. The process flow of ADGE microarray. The control and tester double stranded cDNA are synthesized from RNA and cut with *Taq*I. The control and tester *Taq*I fragments are ligated with the CT and TT adapter, respectively. The adapter-linked control and tester DNA are reassociated through mixing in equivalent amounts, denaturing, followed by annealing. The reassociated DNA is used as template to generate Cy3 labeled control DNA with the CT primer and Cy5 labeled tester DNA with the TT primer. The labeled PCR products are hybridized on a microarray chip.

the templates of the reassociated DNA, the control DNA is amplified by using PCR with the CT primer complementary to the CT adapter while the tester DNA is amplified with the TT primer complementary to the TT adapter. The control and tester PCR products are separated on a gel of high resolution or detected with other methods.

1.1. Principle of Quadratic Magnification of Ratios

Ratio magnification is achieved with the integration of DNA reassociation and polymerase chain reaction (PCR) amplification. DNA reassociation occurs when the control and tester cDNA fragments are mixed in equivalent amounts, denatured, and annealed. DNA reassociation results in the formation of five different duplexes; the control DNA with the CT adapters on both ends, tester DNA with the TT adapters on both ends, hybrid DNA with the CT adapter on one end and the TT adapter on the other end, the end fragment of control DNA with the CT adapter on a single end, and the end fragment of tester DNA with the TT adapter on a single end (**Fig. 1**). The relative amounts of the first three types of duplexes for each gene is theoretically determined by the algebraic formula:

$$(a + b)(a' + b') = aa' + bb' + a'b + ab',$$

where:

a, a' equals the number of sense and antisense strands of the control DNA,
b, b' equals the number of sense and antisense strands of the tester DNA,
aa' equals the number of double strands of the control DNA,
bb' equals the number of double strands of the tester DNA,
a'b, ab' equals the number of double strands of the hybrid DNA.

For example, for a gene overexpressed twofold in tester over control, bb'/aa' = 2. Thus the formula is

$$(a + 2b)(a' + 2b') = aa' + 4bb' + 2a'b + 2ab'.$$

After DNA reassociation, the ratio of bb'/aa' is magnified from 2 to 4. If another gene is down-regulated threefold in tester, aa'/bb' = 3. The formula is

$$(3a + b)(3a' + b') = 9aa' + bb' + 3a'b + 3ab'.$$

Thus, the ratio of aa'/bb' increases from 3 to 9 after DNA reassociation. For a gene with the same transcription level between tester and control, the ratio remains the same after DNA reassociation. For the last two types of duplexes, the hybrid duplex of end fragments cannot be distinguished from the control or tester duplex. Thus, the relative amounts of these two types of duplexes remain the same.

After DNA reassociation, the ratio of control and tester DNA with adapters on the two ends has been magnified quadratically. However, they are not separated from each other or from the hybrid DNA and end fragments. Subsequent PCR is used to separate control DNA from tester DNA. The CT primer complementary to the CT adapter amplifies the control DNA exponentially, and the hybrid and end fragment DNA linearly, because control DNA has the CT adapter on both ends and the hybrid and end fragment DNA have the CT adapter on only one end. The TT primer complementary to the TT adapter amplifies tester DNA exponentially and the hybrid and end fragment DNA linearly (**Fig. 1**). After 20 or more cycles of PCR, the exponentially amplified control or tester DNA is a million times more prevalent than the linearly amplified DNA. Therefore, the ratio of each gene between PCR products with the CT and TT primers is quadratically magnified from the ratio between the control and tester sample. A test experiment showed that the correlation between the detected ratio (y) and the input ratio (x) for ADGE microarray is $y = 1.05x^{1.55}$ with $R^2 = 0.97$ while the correlation is $y = 0.56x + 0.39$ with $R^2 = 0.96$ for standard microarray *(2)*.

1.2. Design of Adapters and Primers

The structure of the adapters and primers is critical to ensure the quadratic magnification of the expression ratios. The basic structure should be the same, although the sequences may differ depending on the selected restriction enzyme and other factors. The CT and TT adapters and primers are designed for the *Taq*I restriction enzyme. The adapters are composed of long and short oligos. The short oligos have the same sequence between CT and TT adapters in order to form hybrid DNA molecules. The complementary region is usually seven nucleotides. If it is too short, the adapters may not be stable. If it is too long, cross priming becomes possible. The length of the unique 5' region between the CT and TT primers should be sufficient to prevent cross priming (at least 10 nucleotides).

The adapters have cohesive ends complementary to *Taq*I. Because the nucleotide T changes to A, the *Taq*I site is not recovered after ligation. The CT and TT primers cover the regions corresponding to the *Taq*I site and complementary to the adapters. If only a portion of transcriptome is needed to amplify in a PCR reaction, selective nucleotides should be added at the 3' end of a primer. The number of selective nucleotides has an inverse relationship with the number of genes that are amplified during each PCR reaction. Four to eight bands were observed on average when four selective nucleotides were used *(1)*. Selective nucleotides are not used in ADGE microarray when all genes are expected to amplify in one PCR reaction. One sequence example of CT, TT adapters, and primers is shown at the bottom of **Fig. 1**.

*Taq*I is optimum for most cases. First, *Taq*I recognizes four nucleotides. In principle, there is a *Taq*I site every 256 nucleotides and two *Taq*I sites every 512 nucleotides. Therefore, the average length of *Taq*I fragments will be 256 base pairs, a good template for PCR and a good probe for chip hybridization. In addition, most genes have two *Taq*I sites to generate the legitimate PCR template. Secondly, *Taq*I generates a cohesive end improving ligation efficiency. If a gene of interest does not have two *Taq*I sites or more, another restriction enzyme which has the similar properties of *Taq*I can be substituted.

1.3. Integration of ADGE and DNA Microarray

The throughput of ADGE is low with agarose gels. DNA microarray technologies are designed to reveal gene expression profiles by simultaneously detecting expression levels on a genomic scale (3). Although pairwise comparison is used in both ADGE and DNA microarray, the ADGE method can be seamlessly combined in series with DNA microarray (**Fig. 1**) (2,4). For coupling ADGE and DNA microarray (hereafter called ADGE microarray) Cy3 and Cy5 dyes are incorporated into the ADGE control and tester DNA, which are in turn hybridized to a DNA chip. The adapters at the DNA fragment ends, facilitate efficient incorporation of Cy3 and Cy5 into DNA templates. Although there is not a selective nucleotide at the 3' end of the CT and TT primers, all genes in the control sample may potentially be amplified with the CT primer in one PCR reaction, and all genes in the tester sample may be amplified with the TT primer in another PCR reaction. Both direct and indirect labeling can be utilized. Using direct labeling, Cy3-dCTP is incorporated into control and Cy5-dCTP into tester DNA during the PCR amplification of the reassociated DNA templates. The use of Cy dyes can be reversed if required. Utilizing indirect labeling, aminoallyl-dUTP is incorporated into control and tester DNA with PCR after the reassociated DNA templates are amplified. The aminoallyl-dUTP labeled DNA templates are in turn coupled with Cy dyes. It is possible to adapt other methods of signal enhancement to the labeling procedure. For example, the 3DNA fluorescent dendrimer probes (Genisphere, Montvale, NJ) can be attached to the CT and TT primers to amplify the control and tester DNAs. Because both strands of the probe DNA are labeled with direct labeling or indirect labeling, the specific activity can be enhanced. The ratio of signal to background was improved in ADGE microarray (2).

1.4. Advantages of ADGE Microarray

When combining ADGE with DNA microarray, the ADGE method is used in sample preprocessing to magnify the ratios of differential gene expression and to amplify the amount of DNA template, while DNA microarray is used as a displaying system to detect differences in gene expression. The combination of ADGE and DNA microarray provides a mutual complementarity of their strengths.

The quadratic magnification of ratios of differential gene expression improves the detection sensitivity. Small changes in gene expression are increased to a level beyond the inherent limit of DNA microarray. Thus, genes with small expression changes could be identified more accurately. For example, the MA plot of ADGE microarray has a wider upward and downward distribution from the central area than that of standard microarray (**Fig. 2**). The ratio magnification is observed over the entire range of spot intensities. The relationship between the detected ratio (y) and the input ratio (x) is $y = 1.05x^{1.55}$ within a 30-fold detected ratio *(2)*.

The PCR amplification of template increases the amount of probe and reduces the requirement for large amounts of starting material. The adapters at the DNA fragment ends facilitate efficient incorporation of Cy3 and Cy5 into DNA templates and enhance signal intensity. Because both strands of the probe DNA are labeled with direct labeling or indirect labeling, the fluorescence intensity is enhanced. In one experiment, 100 ng of total RNA was used to give results showing 6100 out of 10,000 genes with signal intensities higher than background *(4)*. In most cases, 150 ng of total RNA is enough for hybridization on a high density slide, compared with 10 to 20 µg of total RNA in standard microarray and 1 µg with the assistance of T7 promoter amplification *(5)*.

The high throughput of DNA microarray is maintained. Because no selective nucleotide is used on the CT and TT primers, the whole transcriptome may be amplified for the control or tester sample in one PCR reaction. Thus, only one slide hybridization is required for one pair of samples.

2. Materials

1. SuperScript double-stranded cDNA synthesis kit (Invitrogen).
2. *Taq*I (Invitrogen).
3. T4 DNA ligase (Promega).
4. Advantage cDNA polymerase mix (Clontech).
5. QIAquick PCR purification kit (Qiagen).
6. 0.5 *M* EDTA (pH 8.0).
7. Phenol:chloroform:isoamyl alcohol (25:24:1), stored at 4°C.
8. 20 µg/µL glycogen (Sigma), stored at −20°C.
9. 3 *M* sodium acetate (pH 5.2) and 100 m*M* sodium acetate (pH 5.2).
10. 100% ethanol and 70% ethanol.
11. 6X EE buffer: 60 m*M* EPPS, 3 m*M* EDTA (pH 8.0).
12. 3 *M* NaCl.

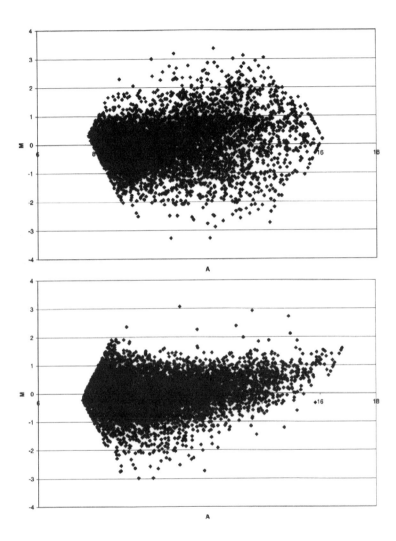

Fig. 2. The MA plots of ADGE microarray (**A**) and standard microarray (**B**). **A** is the average of $\log_2 Cy5$ (the tester sample, the HL60/TLK286 cell line resistant to a prodrug TLK286 *[6]*) and $\log_2 Cy3$ (the control sample, the wild type HL60 cell line), representing intensities of spots. M is the difference of $\log_2 Cy5$ and $\log_2 Cy3$, representing the expression ratios in the power of 2, with positive values for upregulated genes, negative values for downregulated genes and zero for unchanged genes (permission from IUBMB Life).

13. dNTPs/aa-dUTP: a mixture of 10 m*M* dGTP, dATP, dCTP each, 2 m*M* dTTP and 8 m*M* aminoallyl -dUTP (Sigma), stored at −20°C.
14. Cy3 and Cy5 mono-reactive dyes (Amersham). Each pack suspended in 45 µL DMSO and stored at −20°C. Minimize light exposure.
15. 2X coupling buffer: 0.2 *M* sodium bicarbonate (NaHCO₃), pH 9.0.
16. 2X chip hybridization buffer: 10X SSC, 0.2%SDS, 2 µg/µL salmon DNA, 2 µg/µL poly(A), 25X Denhardt's solution, stored at −20°C.
17. Wash buffer I: 2X SSC, 0.1% SDS; wash buffer II: 1X SSC; wash buffer III: 0.2X SSC.

3. Methods

3.1. Synthesis of First-Strand cDNA

1. Take equivalent amounts of control and tester total RNA (in the range of 0.2–10 µg [*see* **Note 1**]) and set up the following reaction:

	Control	Tester
Total RNA	x µL	y µL
Oligo(dT)$_{12-18}$	1 µL	1µL
H₂O (RNase free)	12-x µL	12-y µL

2. Incubate at 70°C for 10 min, chill on ice, then centrifuge briefly.
3. Make the following master mix and add 7µL to each reaction.

5X First-strand buffer	8µL
0.1 *M* DTT	4µL
10 m*M* dNTPs mix	2µL

4. Add 1µL of Superscript II RT to each reaction and mix gently.
5. Incubate 1 h at 42°C (best to work in PCR thermal cycler with hot lid to prevent evaporation), then keep the reactions on ice.

3.2. Synthesis of Second-Strand cDNA

1. Add 130µL of the following second-strand DNA synthesis mixture to each first-strand reaction.

DEPC-treated H₂O	182 µL
10 m*M* dNTPs	6 µL
5X Second-strand buffer	60 µL
E. coli. DNA polymerase	8 µL
E. coli. RNase H	2 µL
E coli. DNA ligase	2 µL

2. Mix well and incubate at 16°C for 2 h, then place on ice.
3. Add 10 µL of 0.5 *M* EDTA (pH 8.0) to each.
4. Add 160 µL of phenol:chloroform:isoamyl alcohol (25:24:1), vortex thoroughly, and centrifuge at room temperature for 5 min at 14,000*g*, transfer the upper, aqueous layer to a fresh 1.5 mL Eppendorf tube.

5. Add 1 µL of 20 µg/µL glycogen, then add 20 µL of 3 *M* sodium acetate (pH 5.2), add 300 µL of absolute ethanol (100%), vortex the mixtures, centrifuge at 4°C for 20 min at 14,000*g*, remove the supernatant and discard.

6. Add 500 µL of 70% ethanol to the pellet, centrifuge 5 min at 14,000*g*, remove the supernatant and discard, dry the cDNA at 37°C for 10 min to evaporate the residual ethanol.

7. Resuspend each cDNA pellet in 25 µL of ddH$_2$O.

3.3. Taq*I* Digestion

1. Take equivalent amounts of control and tester cDNA and set up the following reactions:

	Control	Tester
cDNA	25 µL	25 µL
Restriction buffer 2	3 µL	3 µL
*Taq*I	2 µL	2 µL

2. Mix and incubate at 65°C for 2 h.

3.4. Ligation of the CT and TT Adapters

1. Set up the following ligation reaction:

	Control	Tester
*Taq*I fragments	30 µL	30 µL
10X T4 ligase buffer	4 µL	4 µL
60 µ*M* CT Adapter	4 µL	
60 µ*M* TT Adapter		4 µL
T4 ligase	2 µL	2 µL

2. Incubate overnight at 14°C.

3.5. Reassociation of Control and Tester DNA

1. Set up the following reaction, x is usually from 5 to 15 µL (the amounts of control and tester DNA should be equivalent [*see* **Note 2**]).

CT adapter-control cDNA	x µL
TT adapter-tester cDNA	x µL
6X EE buffer	2x µL

2. Mix well, denature at 95°C for 5 min, chill on ice immediately.

3. Add 2X µL of 3 M NaCl, mix well, incubate at 68°C overnight.

4. Purify the reactions with QIAquick PCR purification kit (Qiagen) and elute into 50 µL ddH$_2$O, dry down the volume if a small amount of starting material is used.

3.6. Labeling Probes With PCR

1. Set up the following PCR reaction to amplify the DNA template. x could be from 1 to 42 µL, depending on the amount of starting material. Two or three such reactions should be set up for stronger signal (*see* **Note 3**).

	Control	Tester
Reassociated DNA	x μL	x μL
10X PCR buffer	5 μL	5 μL
10 m*M* dNTPs	1 μL	1 μL
CT primer	1 μL	
TT primer		1 μL
cDNA polymerase	1 μL	1 μL
ddH$_2$O	42-x μL	42-x μL

The PCR cycles: 72°C for 5 min (for filling in the ends), 94°C for 1 min, 25 cycles of 94°C for 30 s, 66°C for 30 s, 72°C for 90 s, then 72°C for 5 min, then stored at 4°C.

2. Purify the PCR products with QIAquick PCR purification kit and elute into 42 μL ddH$_2$O.

3. Set up the following PCR reaction to incorporate aa-dUTP:

	Control	Tester
DNA templates	42 μL	42 μL
10X PCR buffer	5 μL	5 μL
dNTPs/aa-dUPT	1 μL	1 μL
CT primer	1 μL	
TT primer		1 μL
cDNA polymerase	1 μL	1 μL

The PCR cycles: 94°C for 1 min, 6 cycles of 94°C for 30 s, 64°C for 30 s, 72°C for 90 s, then 72°C for 1 min, 4°C for storage.

4. Add 1 μL of 20 μg/μL glycogen and 10 μL of 3 *M* sodium acetate (pH5.2), add 100 μL of absolute ethanol (100%), vortex the mixtures, centrifuge at 4°C for 20 min at 14,000*g*, remove the supernatant and discard.

5. Add 500 μL of 70% ethanol to the pellet, centrifuge 5 min at 14,000*g*, remove the supernatant and discard, dry the cDNA at 37°C for 10 min to evaporate the residual ethanol.

6. Resuspend the DNA pellet in 5 μL 2X coupling buffer, add 5 μL Cy3 mono-reactive dye to the control DNA and 5 μL Cy5 monoreactive dye to the tester DNA, incubate at room temperature in the dark for 1 h.

7. Add 50 μL of 100 m*M* sodium acetate (pH 5.2) to each reaction, proceed to PCR purification with QIAquick PCR purification kit, elute two times with 40 μL ddH$_2$O (prewarmed at 42°C) each, and vacuum dry to 7 μL.

3.7. Chip Hybridization

1. Add 7 μL of 2X chip hybridization buffer (prewarmed at 42°C) to the control and tester probes and mix well.

2. Denature at 95°C for 5 min, chill on ice for 5 min and incubate at 42°C for 10 min.

3. Mix the denatured control and tester probes together, load onto a microarray chip.

4. Assemble the hybridization chamber with the microarray chip and incubate in a water bath at 58°C overnight.

3.8. Chip Washing and Scanning

1. Disassemble the hybridization chamber, submerge in wash buffer I in a slide washing jar and let the cover slip slide out.
2. Transfer the slide to a fresh wash buffer I and shake for 5 min.
3. Transfer the slide to wash buffer II and shake for 5 min.
4. Transfer the slide to wash buffer III and shake for 5 min.
5. Dry the slide by centrifuging at 700*g* for 5 min and scan with an array scanner using the Cy3 and Cy5 channels.

4. Notes

1. Starting material: It is unnecessary to use mRNA. The preferred amount of total RNA is 5–10 μg though as little as 0.2 μg of total RNA is enough for one chip hybridization. If a small amount of total RNA is used, the eluate after DNA reassociation should be concentrated by drying down the volume.
2. Monitoring equivalent amounts of control and tester DNA: DNA reassociation requires equivalent amounts of control and tester DNA. One way of checking the equivalence between control and tester cDNA is detecting actin levels with actin ADGE primers. Actin ADGE primers are the CT and TT primers with selective nucleotides complementary to the actin gene. One μL aliquot of the control and tester DNA is diluted 10 and 100 times after adapter ligation. The diluted DNA is used as template for PCR with the actin ADGE primers. The actin levels are compared between the control and tester. The volumes of the control and tester for DNA reassociation are adjusted to make equivalence.
3. Labeling probe: Depending on the printed area of microarray chip, two or three labeling reactions for the control and tester are needed for one slide hybridization. Both indirect (**Subheading 3.6.**) and direct (**Subheading 4.4.**) PCR labeling work well. Direct labeling usually gives a stronger signal and has a simpler procedure but requires expensive reagents, Cy3- and Cy5-dCTP.
4. Direct PCR labeling: Set up the PCR reaction with dNTPs (10m*M* dGTP, dATP, dTTP each, 6m*M* dCTP). x could be from 1 to 42 μL, depending on the amount of starting material. Two or three such reactions should be set up for stronger signal (*see* **Note 3**).

	Control	Tester
Reassociated DNA	x μL	x μL
10X PCR buffer	5 μL	5 μL
dNTPs	1 μL	1 μL
Cy3-dCTP	4 μL	
CT primer	2 μL	
Cy5-dCTP		4 μL
TT primer		2 μL
cDNA polymerase	1 μL	1 μL
ddH2O	37-x μL	37-x μL

The PCR cycles: 72°C for 5 min, 94°C for 1 min, 30 cycles of 94°C for 30 s, 62°C for 30 s, 72°C for 90 s, then 72°C for 5 min, 4°C for storage.
PCR purification with QIAquick PCR purification kit, elute two times with 40 μL ddH2O (prewarmed at 42°C) each, and vacuum dry down to 5 μL.

Table 1
Common Problems and Troubleshooting Guide

Symptom	Cause	Solution
Weak signal of spots	• Too little probe • Inefficient incorporation of Cy dyes	• Make more probes • Improve elution efficiency with prewarmed ddH$_2$O • Completely dissolve DNA, mix well in coupling reaction
High background	• Not enough blocking DNA • Low hybridization temperature • Not enough washes	• Increase amount of salmon DNA and poly(A) in hybridization buffer • Raise the hybridization temperature • Extend washing time, especially in wash buffer 3
Systematically skewed ratios between control and tester after normalization	• Non-equivalent amounts of control and tester DNA for DNA reassociation	• Check the equivalence of actin level as in section 4.2

5. Common problems: weak signal, high background, and skewed ratios are among common problems. The guideline of troubleshooting is provided (**Table 1**).

References

1. Chen, Z. J., Shen, H., and Tew, K. D. (2001) Gene expression profiling using a novel method: amplified differential gene expression (ADGE). *Nucleic Acids Res.* **29,** e46.
2. Chen, Z. J., Gaté, L., Davis, W. Jr., Ile, K. E., and Tew, K. D. (2003) Improving gene expression profiling with the combination of DNA microarray and amplified differential gene expression (ADGE). *BMC Genomics* **4,** 28.
3. Schena, M., Shalon, D., Davis, R. W., and Brown, P. O. (1995) Quantitative monitoring of gene expression patterns with a complementary DNA microarray. *Science* **270,** 467–470.
4. Chen, Z. J., Gaté, L., Davis, W. Jr., Ile, K. E., and Tew, K. D. (2003) Integration of amplified differential gene expression (ADGE) and DNA microarrray. *IUBMB* **54,** 1–4.
5. Pabon, C., Modrusan, Z., Ruvolo, M. V., et al. (2001) Optimized T7 amplification system for microarray analysis. *BioTechniques* **31,** 874–879.
6. Rosario, L. A., O'Brien, M. L., Henderson, C. J., Wolf, C. R., and Tew, K. D. (2000) Cellular response to a glutathione S-transferase P1-1 activated prodrug. *Mol. Pharmacol.* **58,** 167–174.

8

Suppression Subtractive Hybridization

Denis V. Rebrikov, Sejal M. Desai,
Paul D. Siebert, and Sergey A. Lukyanov

Summary

Suppression subtractive hybridization (SSH) is a widely used method for separating DNA molecules that distinguish two closely related DNA samples. Two of the main SSH applications are cDNA subtraction and genomic DNA subtraction. In fact, SSH is one of the most powerful and popular methods for generating subtracted cDNA or genomic DNA libraries. The SSH method is based on a suppression PCR effect and combines normalization and subtraction in a single procedure. The normalization step equalizes the abundance of DNA fragments within the target population, and the subtraction step excludes sequences that are common to the populations being compared. This dramatically increases the probability of obtaining low-abundance differentially expressed cDNA or genomic DNA fragments, and simplifies analysis of the subtracted library. In our hands, the SSH technique has enriched over 1000-fold for rare sequences in a single round of subtractive hybridization.

Key Words: cDNA, mRNA, normalization, subtractive hybridization

1. Introduction

Suppression subtractive hybridization (SSH) is a widely used method for separating DNA molecules that distinguish two closely related DNA samples. Among SSH's many applications (**Table 1**), are cDNA and genomic DNA subtraction. In fact, SSH is one of the most powerful and popular methods for generating subtracted cDNA or genomic DNA libraries *(1–4)*.

The SSH method is based on a suppression polymerase chain reaction (PCR) effect *(5,6)* and combines normalization and subtraction in a single procedure *(2)*. The normalization step equalizes the abundance of DNA fragments within

From: *Methods in Molecular Biology, Vol. 258: Gene Expression Profiling: Methods and Protocols*
Edited by: R. A. Shimkets © Humana Press Inc., Totowa, NJ

Table 1
Applications of SSH and MOS

Sample type	Sequences to be found during SSH
cDNA	Differentially expressed sequences
Bacterial genome	Unique and amplified sequences that are present in one bacterial genome, but are absent in another.
Eukaryotic genome	SSH and MOS combination required. Mainly the repeated sequences, mobile elements and extrachromosomal elements that are present in one eukariotic genome, but are absent in another.

the target population, and the subtraction step excludes sequences that are common to the compared populations. This dramatically increases the probability of obtaining low-abundance differentially expressed cDNA or genomic DNA fragments, and it simplifies analysis of the subtracted library. In our lab, the SSH technique has enriched more than 1000-fold rare sequences in a single round of subtractive hybridization *(2,4,7)*.

Nevertheless, in practice, not all differentially expressed genes are equally enriched by SSH. The level of enrichment of a particular cDNA depends on its original abundance, the ratio of its concentration in the samples being subtracted, and the number of other differentially expressed genes *(8)*. Other factors, such as the complexity of a starting material, hybridization time, and ratio of two samples being subtracted, play a very important role in SSH's success in a given application *(8)*. For instance, the high complexity of mammalian genomic DNA makes SSH application very difficult. Likewise, some cDNA subtractions are also very challenging because of the nature of the starting samples *(8)*. Subtracted libraries generated using complex samples may contain very high background. An especially challenging problem is the inclusion of so-called "false positive" clones that generate a differential signal in a primary screening procedure, but are not confirmed by subsequent detailed analysis. To overcome this problem, a simple procedure called mirror orientation selection (MOS) can be used to substantially decrease the number of background clones *(9)*.

In this chapter, we describe the SSH technique for generating subtracted cDNA or genomic DNA libraries. A detailed protocol for cDNA synthesis, subtractive hybridization, PCR amplification, library generation, and differential screening analysis is provided. We also describe the MOS procedure that substantially decreases the number of background clones in SSH-generated libraries. Finally, we show an example of SSH- and MOS-subtracted library.

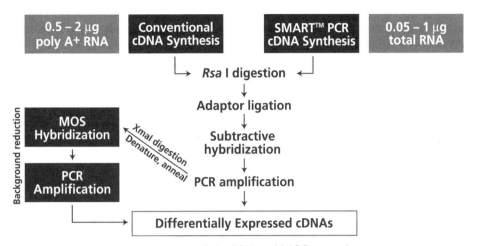

Fig. 1. Overview of the SSH and MOS procedures.

1.1. The Principle of SSH and MOS

Figure 1 presents a brief overview of the SSH and MOS procedures. SSH includes several steps. First, cDNA is synthesized from the two types of tissues or cell populations being compared. The cDNA population in which specific transcripts are to be found is called *tester cDNA,* and the reference cDNA population is called *driver cDNA.* For cDNA synthesis, the conventional method described by Gubler and Hoffman *(10)* can be used. If enough poly(A)+ RNA is not available, the Switch Mechanism at the 5' end of RNA Templates (SMART^{TM}) amplification technology (BD Biosciences Clontech) can be used to preamplify high-quality cDNA from total RNA *(11).* In the second step, tester and driver cDNAs are digested with a four-base-cutting restriction enzyme that yields blunt ends, such as *Rsa* I. The tester cDNA is then subdivided into two portions, and each is ligated to a different double-stranded (ds) adaptor (adaptors 1 [Ad1] and 2R [Ad2R]). The ends of the adaptors are not phosphorylated, so only one strand of each adaptor becomes covalently attached to the 5' ends of the cDNAs.

The molecular events that occur during subtractive hybridization and selective amplification of differentially expressed genes are illustrated in **Fig. 2**. In the first hybridization, an excess of driver cDNA is added to each sample of tester cDNA. The samples are then heat-denatured and allowed to anneal. **Figure 2A** shows the type A, B, C, and D molecules generated in each sample. Type A molecules, which represent single-stranded (ss) tester molecules, include equal concentrations of high- and low-abundance sequences because reannealing is faster for more abundant molecules owing to the second-order kinetics of hybrid-

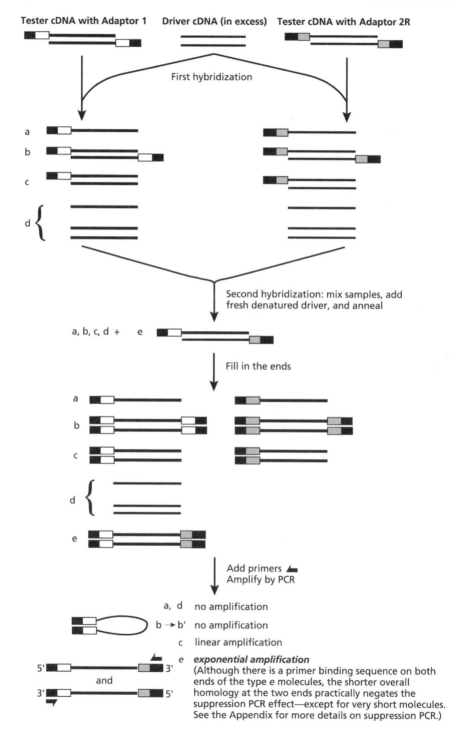

ization *(12)*. At the same time, the population of type A molecules is significantly enriched for differentially expressed sequences because common non-target cDNAs form type C molecules with the driver. During the second hybridization, the two primary hybridization samples are mixed together in the presence of fresh denatured driver. Type A cDNAs from each tester sample are now able to associate and form type B, type C, and new type E hybrids. Type E hybrids are ds tester molecules with different ss ends that correspond to Ad1 and Ad2R. Freshly denatured driver cDNA is added to further enrich fraction E for differentially expressed sequences.

The entire population of molecules is then subjected to two rounds of PCR to selectively amplify the differentially expressed sequences. Prior to the first cycle of primary PCR, the adaptor ends are filled in, creating the complementary primer binding sites needed for amplification. Type A and D molecules lack primer annealing sites and cannot be amplified. Type B molecules form a pan handle-like structure that suppresses amplification *(5,6)*. Type C molecules have only one primer annealing site and can only be amplified linearly. Only type E molecules, which have two different primer annealing sites, can be amplified exponentially. Differentially expressed sequences are greatly enriched in type E fraction, and therefore in the subtracted cDNA pool. This method does not involve the physical separation of single stranded and double stranded molecules although suppression PCR prevents undesirable amplification during target molecules enrichment.

When there is a high background in the SSH-generated subtracted library, MOS can be used to reduce the background significantly. The MOS technique is based on the rationale that, after PCR amplification during SSH, each species of background molecule has only one orientation relative to the adaptor sequences. This directionality corresponds to the orientation of the progenitor molecule. On the contrary, the target DNA fragments are involved in PCR amplification owing to efficient enrichment in the SSH procedure. As a result, each specific sequence has many progenitors and is represented by both sequence orientations *(9)*. The procedure includes removing adaptor 1 (represented by adaptor 1-derived primer NP1 in secondary PCR of SSH, **Fig. 3**) by restriction endonuclease (*Xma*I in this description), heat-denaturation and reannealing of the SSH sample (**Fig. 3**). Some of the newly formed hybrids from target DNAs bear

Fig. 2. (*Opposite page*) Schematic diagram of the suppression subtractive hybridization procedure. Solid boxes represent the outer part of the adaptor Ad1 and Ad2R, and correspond to the polymerase chain reaction (PCR) primer 1 (P1) sequence. Clear boxes represent the inner part of adaptor Ad1 and correspond to nested PCR primer 1 (NP1). Shaded boxes represent the inner part of adaptor Ad2R and correspond to nested PCR primer 2R (NP2R).

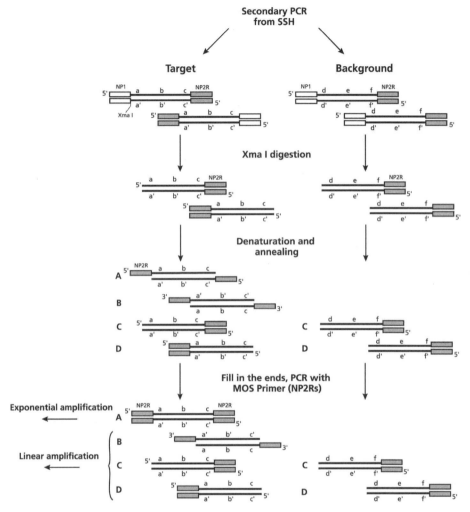

Fig. 3. Schematic diagram of the mirror orientation selection (MOS) procedure. The method is based on the assumption that each background molecule has only one orientation with respect to the Ad1 (represented by NP1) and Ad2R (represented by NP2R) adaptors used in suppression subtractive hybridization (SSH), whereas truly differentially expressed target cDNA fragments are represented by both sequence orientations. MOS polymerase chain reaction primer is a shorter primer sequence of the NP2R primer used in SSH.

adaptor 2R (represented by adaptor 2R-derived primer NP2R in secondary PCR of SSH, **Fig. 3**) at both termini. These molecules are generated as a result of hybridization of molecules with "mirror" orientation of adaptors 1 and 2. Thus, they can only be derived from target DNA fraction. Next, the 3' ends are filled in and PCR with primers corresponding to NP2R (also called NP2Rs or MOS PCR primer) is performed. In this PCR only molecules bearing NP2R at both termini can be amplified exponentially. Thus, the final PCR product is enriched for target sequences.

2. Materials

2.1. Oligonucleotides

The following oligonucleotides are used at a concentration of 10 μM. Whenever possible, oligonucleotides should be gel-purified.

1. cDNA synthesis primers: 5'-TTTTGTACAAGCTT$_{30}$-3'
2. Ad1:
   5'-CTAATACGACTCACTATAGGGCTCGAGCGGCCGCCCGGGCAGGT-3'
   3'-GGCCCGTCCA-5'
   Ad2R:
   5'-CTAATACGACTCACTATAGGGCAGCGTGGTCGCGGCCGAGGT-3'
   3'-GCCGGCTCCA-5'
3. PCR primer 1 (P1): 5'-CTAATACGACTCACTATAGGGC-3'
4. Nested primer 1 (NP1): 5'-TCGAGCGGCCGCCCGGGCAGGT -3'
5. Nested primer 2R (NP2R): 5'-AGCGTGGTCGCGGCCGAGGT-3'
6. MOS PCR primer (NP2Rs): 5'-GGTCGCGGCCGAGGT-3'
7. Blocking solution: A mixture of the cDNA synthesis primer, nested primers (NP1 and NP2R), and their respective complementary oligonucleotides (2 mg/mL each).

2.2. Buffers and Enzymes

All chemical reagents were obtained from Sigma Chemical (St. Louis, MO)

1. AMV reverse transcriptase (20 U/μL; Life Technologies, Gaithersburg, MD).
2. 5X First-strand buffer: 250 mM Tris-HCl, pH 8.5, 40 mM MgCl$_2$, 150 mM KCl, and 5 mM Dithiothreitol.
3. 20X Second-strand enzyme cocktail: DNA polymerase I, 6 U/μL, New England Biolabs, Beverly, MA.
4. RNase H, 0.25 U/μL, Epicentre Technologies, Madison, WI.
5. *E. coli* DNA ligase, 1.2 U/μL, New England Biolabs.
6. 5X Second-strand buffer: 500 mM KCl, 50 mM ammonium sulfate, 25 mM MgCl$_2$, 0.75 mM b-NAD, 100 mM Tris-HCl, pH 7.5, 0.25 mg/ml BSA.
7. T4 DNA polymerase (3 U/μL, New England Biolabs).
8. 10X *Rsa* I restriction buffer: 100 mM Bis-Tris Propane/HCl, pH 7.0, 100 mM MgCl$_2$ and 1 mM Dithiothreitol (DTT).
9. *Rsa* I (10 U/mL, New England Biolabs).
10. T4 DNA ligase (400 U/μL: contains 3 mM ATP, New England Biolabs).
11. 5X DNA ligation buffer: 250 mM Tris-HCl, pH 7.8, 50 mM MgCl$_2$, 10 mM DTT, 0.25 mg/mL BSA.

12. 4X Hybridization buffer: 4 M NaCl, 200 mM HEPES, pH 8.3, 4 mM cetyltrimethyl ammonium bromide (CTAB).
13. Dilution buffer: 20 mM HEPES-HCl, pH 8.3, 50 mM NaCl, 0.2 mM EDTA.
14. Advantage cDNA PCR Mix (BD Biosciences Clontech, Palo Alto, CA). This mix contains a mixture of KlenTaq-1 and DeepVent DNA polymerases (New England Bio Labs, Beverly, MA) and TaqStart Antibody (BD Biosciences Clontech); 10X reaction buffer (40 mM Tricine-KOH (pH 9.2 at 22°C), 3.5 mM Mg(OAc)$_2$, 10 mM KOAc, 75 mg/mL BSA). The TaqStart Antibody provides automatic hot start PCR *(13)*. Alternatively, *Taq* DNA polymerase can be used alone, but five additional thermal cycles will be needed in both the primary and secondary PCR, and the additional cycles may cause higher background. If the Advantage cDNA PCR Mix is not used, manual hot start or hot start with wax beads is strongly recommended to reduce nonspecific DNA synthesis.
15. 10 mM each dNTP (Amersham Pharmacia Biotech, Piscataway, NJ).
16. 20X EDTA/glycogen mix: 0.2 M EDTA, 1 mg/mL glycogen.
17. 4 M NH$_4$OAc, TN buffer: 10 mM Tris-HCl, 10 mM NaCl.
18. ExpressHyb™ Hybridization Solution (BD Biosciences Clontech).
19. Sterile H$_2$O.

Please note that the cycling parameters in this protocol have been optimized using the MJ Research PTC-200 DNA Thermal Cycler. For a different type of thermal cycler, the cycling parameters must be optimized for that machine. It is not possible to use this protocol with water-bath thermal cyclers because there is no PCR suppression effect there.

We recommend performing subtractions in both directions for each DNA pair being compared. This forward- and reverse-subtracted DNA may be useful for differential screening of the resulting subtracted libraries. We also recommend performing self-subtractions (with both tester and driver prepared from the same DNA sample) as a control experiment for fast examination of subtraction efficiency (*see* **Note 1**). For models such as RNA/DNA injections, or viral infections, it is extremely important to add appropriate DNA into driver sample (*see* **Note 2**).

3. Methods

3.1. Preparation of Subtracted cDNA or Genomic DNA Library

3.1.1. RNA and DNA Isolation

Two µg of genomic DNA or RNA is required per subtraction. Most commonly used methods for isolation of RNA and genomic DNA are appropriate for subtraction experiments *(14–17)*. Nevertheless, the quality of DNA or RNA is very important for successful experiment. Whenever possible, samples being compared should be purified side-by-side utilizing the same reagents and protocol. Alternatively, commercially available kits from different vendors can be used for RNA and DNA isolation.

If genomic DNA is used as a starting material, the next step is *Rsa* I digestion (**Subheading 3.1.3.**). If RNA is used as a starting material, the next step is cDNA synthesis (**Subheading 3.1.2.**).

Note: For simplicity, the term "cDNA" will be used throughout the protocol, but the protocol is suitable for genomic DNA subtraction without any changes in the amount of any reagents required to perform subtraction.

3.1.2. cDNA Synthesis

There are two steps involved in cDNA synthesis: first-strand cDNA synthesis and second-strand cDNA synthesis. During first-strand cDNA synthesis, AMV reverse transcriptase synthesizes cDNA using poly(A)+ RNA as a template. During second-strand cDNA synthesis, DNA polymerase I uses first-strand cDNA as a template. The following protocol is recommended for generating a subtracted library from 2 μg of poly(A)+ RNA. If enough poly(A)+ RNA is not available, the switch mechanism at the 5' end of RNA Templates (SMART™) amplification technology (BD Biosciences Clontech) can be used to preamplify high-quality cDNA from total RNA *(11)* (*see* **Note 3**).

3.1.2.1. FIRST-STRAND CDNA SYNTHESIS

Perform this procedure individually with each tester and driver poly(A)+ RNA sample.

1. For each tester and driver sample, combine the following components in a sterile 0.5-mL microcentrifuge tube (do not use a polystyrene tube).
    poly(A)+ RNA (2 μg) to 2–4 μL
    cDNA synthesis primer (10 μ*M*) to 1 μL
    If needed, add sterile H_2O to a final volume of 5 μL.
2. Incubate the tubes at 70°C in a thermal cycler for 2 min.
3. Cool at room temperature for 2 min and briefly centrifuge using a PicoFuge® microcentrifuge (Stratagene, La Jolla, CA) at maximum rotation speed (5000*g*).
4. Add the following to each reaction tube:
    2 μL 5X first-strand buffer
    1 μL dNTP mixture (10 mM each)
    0.5 μL sterile H2O
    (Optional: To monitor the progress of cDNA synthesis, dilute 0.5 μL of P^{32}-labeled dCTP (10 mCi/mL, 3000 Ci/mmol) with 9 μL of H_2O and replace the H_2O above with 1 μL of the diluted label.)
    0.5 μL 0.1M DTT
    1 μL AMV reverse transcriptase (20 U/μL)
5. Gently vortex and briefly centrifuge the tubes.
6. Incubate the tubes at 42°C for 1.5 h in an air incubator.

7. Place the tubes on ice to terminate first-strand cDNA synthesis and immediately proceed to second-strand cDNA synthesis.

3.1.2.2. SECOND-STRAND cDNA SYNTHESIS

1. Add the following components (previously cooled on ice) to the first-strand synthesis reaction tubes:

    48.4 µL Sterile H$_2$O

    16.0 µL 5X Second-strand buffer

    1.6 µL dNTP mix (10 m*M*)

    4.0 µL 20X Second-strand enzyme cocktail

2. Mix the contents and briefly spin the tubes. The final volume should be 80µL.
3. Incubate the tubes at 16°C (water bath or thermal cycler) for 2 h.
4. Add 2 µL (6 U) of T4 DNA polymerase. Mix contents well.
5. Incubate the tube at 16°C for 30 min in a water bath or a thermal cycler.
6. Add 4 µL of 0.2 *M* EDTA to terminate second-strand synthesis.
7. Perform phenol:chloroform extraction and ethanol precipitation (*see* **Note 4**).
8. Dissolve pellet in 50 µL of TN buffer.
9. Transfer 6 µL to a fresh microcentrifuge tube. Store this sample at −20°C until after *Rsa* I digestion. This sample will be used for agarose gel electrophoresis to estimate yield and size range of the ds cDNA synthesized products.

3.1.2.3. *RSA* I DIGESTION

Perform the following procedure with each experimental ds tester and driver cDNA. This step generates shorter, blunt-ended ds cDNA fragments optimal for subtractive hybridization.

1. Add the following reagents into the tube from **Subheading 3.1.2.2., step 8**:

    43.5 µL ds cDNA

    5.0 µL 10X *Rsa* I restriction buffer

    1.5 µL *Rsa* I (10 U/µL)

2. Mix and incubate at 37°C for 2–4 h.
3. Use 5 µL of the digest mixture and analyze on a 2% agarose gel along with undigested cDNA (**Subheading 3.1.2.2., step 9** or **Subheading 3.1.1.** for genomic DNA) to analyze the efficiency of *Rsa* I digestion.

    Note: continue the digestion during electrophoresis and terminate the reaction only after you are satisfied with the results of the analysis.
4. Add 2.5 µL of 0.2 *M* EDTA to terminate the reaction.
5. Perform phenol:chloroform extraction and ethanol precipitation (*see* **Note 4–6**).
6. Dissolve each pellet in 6 µL of TN buffer (*see* **Note 7**) and store at −20°C.

    Driver cDNA preparation is now complete.

3.1.3. Adaptor Ligation

It is strongly recommended that you perform subtractions in both directions for each tester/driver cDNA pair. Forward subtraction is designed to enrich differentially expressed transcripts present in tester but not in driver; reverse subtraction is designed to enrich differentially expressed sequences present in driver but not in tester. The availability of such forward- and reverse-subtracted cDNAs will be useful for differential screening of the resulting subtracted tester cDNA library (*see* **Subheading 3.4.**).

The tester cDNAs are ligated separately to Ad1 (Tester 1-1 and 2-1) and Ad2R (Tester 1-2 and 2-2). It is highly recommended that a third ligation of both adaptors 1 and 2R to the tester cDNAs (unsubtracted tester control 1-c and 2-c) be performed and used as a negative control for subtraction. Please note that the adaptors are not ligated to the driver cDNA.

1. Dilute 1 µL of each *Rsa* I-digested tester cDNA from the above section with 5 µL of TN buffer.
2. Prepare a master ligation mix of the following components for each reaction:
   3 µL Sterile H_2O
   2 µL 5X Ligation buffer
   1 µL T4 DNA ligase (400 U/µL)
   Please note that ATP required for ligation is in the T4 DNA ligase (3 mM initial, 300 µM final.
3. For each tester cDNA mixture, combine the following reagents in a 0.5-mL microcentrifuge tube in the order shown. Pipet the solution up and down to mix thoroughly.

Tube No.:	1	2
Component	Tester 1-1 (µL)	Tester 1-2 (µL)
Diluted tester cDNA	2	2
Adaptor Ad1 (10 µM)	2	—
Adaptor Ad2R (10 µM)	—	2
Master ligation mix	6	6
Final volume	10	10

4. In a fresh microcentrifuge tube, mix 2 µL of Tester 1-1 and 2 µL of Tester 1-2. This is your unsubtracted tester control 1-c. Do the same for each tester cDNA sample. After ligation, approximately one-third of the cDNA molecules in each unsubtracted tester control tube will have two different adaptors on their ends, suitable for exponential PCR amplification with adaptor-derived primers.
5. Centrifuge the tubes briefly and incubate at 16°C overnight.
6. Stop the ligation reaction by adding 1 µL of 0.2 M EDTA.
7. Heat samples at 72°C for 5 min to inactivate the ligase.
8. Briefly centrifuge the tubes. Remove 1 µL from each unsubtracted tester control (1-c, 2-c ...) and dilute into 1 mL of H_2O. These samples will be used for PCR amplification (**Subheading 3.1.6.**).

Preparation of your experimental Adaptor-Ligated Tester cDNAs 1-1 and 1–2 is now complete.

Perform ligation efficiency test before proceeding to the next section (*see* **Note 8**).

3.1.5. Subtractive Hybridization

3.1.5.1. FIRST HYBRIDIZATION

1. For each tester sample, combine the reagents in the following order:

Component	Hybridization 1.1 (µL)	Hybridization 1.2 (µL)
Rsa I-digested driver cDNA (**Subheading 3.1.3.**, **step 7**)	1.5	1.5
Ad1-ligated Tester 1-1 (**Subheading 3.1.4.**, **step 8**)	1.5	—
Ad2R-ligated Tester 1-2 (**Subheading 3.1.4.**, **step 5**)	—	1.5
4X Hybridization buffer	1.0	1.0
Final volume	4.0	4.0

2. Overlay samples with one drop of mineral oil and centrifuge briefly.
3. Incubate samples in a thermal cycler at 98°C for 1.5 min.
   Incubate samples at 68°C for 8 h (*see* **Note 9**) and then proceed immediately to the second hybridization (*see* **Note 21**).

3.1.5.2. SECOND HYBRIDIZATION

1. Repeat the following steps for each experimental driver cDNA.
   Add the following reagents into a sterile 0.5-µL microcentrifuge tube:
   1 µL Driver cDNA (**Subheading 3.1.3**, **step 7**)
   1µL 4X hybridization buffer
   2 µL Sterile H_2O
2. Place 1 µL of this mixture in a 0.5-mL microcentrifuge tube and overlay it with one drop of mineral oil.
3. Incubate in a thermal cycler at 98°C for 1.5 min (*see* **Note 10**).
4. Remove the tube of freshly denatured driver from the thermal cycler (*see* **Note 11**).
5. To the tube of freshly denatured driver cDNA, add hybridized sample 1.1 and hybridized sample 1.2 (from first hybridization) in that order. This ensures that the two hybridization samples are mixed only in the presence of excess driver cDNA.
6. Incubate the hybridization reaction at 68°C overnight.
7. Add 100 µL of dilution buffer to the tube and mix well by pipeting.
8. Incubate in a thermal cycler at 72°C for 7 min.
9. Store hybridization solution at −20°C (*see* **Note 12**).

3.1.5.3. PCR AMPLIFICATION

Differentially presented DNAs are selectively amplified during the reactions described in this section. Each experiment should have at least four reactions: subtracted tester cDNAs, unsubtracted tester control (1-c), reverse-subtracted tester cDNAs, and unsubtracted driver control for the reverse subtraction (2-c).

3.1.5.4. PRIMARY PCR

1. Place a 1 µL aliquot of each diluted cDNA sample (i.e., each subtracted sample from **Subheading 3.1.5.3.**, **step 8**, and the corresponding diluted unsubtracted tester control

from **Subheading 3.1.4.**, **step 8**) into an appropriately labeled tube (*see* **Note 12**).

2. Prepare a master mix for all of the primary PCR tubes plus one additional tube. For each reaction combine the reagents in the order shown:

Reagent	Amount per reaction (μL)
Sterile H_2O	19.5
10X PCR reaction buffer	2.5
dNTP mix (10 m*M*)	0.5
PCR primer P1 (10 μ*M*)	1.0
50X Advantage cDNA PCR Mix	0.5
Total volume	24.0

3. Place 24 μL aliquot of master mix into each reaction tube prepared in **step 1**.
4. Overlay with 50 μL of mineral oil. Skip this step if an oil-free thermal cycler is used.
5. Incubate the reaction mixture in a thermal cycler at 75°C for 5 min to extend the adaptors (*see* **Note 13**). Do not remove the samples from the thermal cycler.
6. Immediately commence 26 cycles of:

   95°C   10 s
   66°C   10 s
   72°C   1.5 min

7. Analyze 4 μL from each tube on a 2% agarose/EtBr gel run in 1X TAE buffer (*see* **Note 14 and 15**).

3.1.5.5. SECONDARY PCR

1. Dilute 2 μL of each primary PCR mixture in 38 μL of H_2O.
2. Place 1 μL aliquot of each diluted primary PCR product mixture from **step 1** into an appropriately labeled tube.
3. Prepare a master mix for the secondary PCR samples plus one additional reaction by combining the reagents in the following order:

Reagent	Amount per reaction (μL)
Sterile H_2O	18.5
10X PCR reaction buffer	2.5
Nested PCR primer NP1 (10 μ*M*)	1.0
Nested PCR primer NP2R (10 μ*M*)	1.0
dNTP mix (10 m*M*)	0.5
50X Advantage cDNA PCR mix	0.5
Total volume	24.0

4. Place 24 μL aliquot of master mix into each reaction tube from **step 2**.
5. Overlay with one drop of mineral oil. Skip this step if an oil-free thermal cycler is used.
6. Immediately commence 10–12 cycles of:

   95°C   10 s
   68°C   10 s
   72°C   1.5 min

7. Analyze 4 μL from each reaction on a 2% agarose/EtBr gel.
8. Store PCR products at −20°C. This PCR product is now enriched for differentially presented DNAs.

At this point if you are not going to perform MOS, please go to **Subheading 3.3.** (Cloning of subtracted library) in this method section.

3.2. Mirror Orientation Selection (MOS)

The major drawback of SSH is the presence of background clones that represent nondifferentially expressed DNA species in the subtracted libraries. In some difficult cases, the number of background clones may considerably exceed the number of target clones. To overcome this problem, we recommend MOS— a simple procedure that substantially decreases the number of background clones in the libraries generated by SSH (*see* **Note 16**).

We recommend the use of MOS in the following cases:

- If the percentage of differentially expressed clones found during differential screening is very low (for example, 1–5%). The MOS procedure can increase the number of differential clones up to 10-fold.
- If most of the differentially expressed clones found are false positive clones (i.e., clones that appear to be differentially expressed in the differential screening procedure, but turn out not to be differentially expressed in the Northern blot or reverse transcriptase [RT]-PCR analysis). The MOS procedure decreases the portion of false positive clones by several fold.
- If the primary PCR in SSH requires more than 30 cycles (but no more than 36 cycles, *see* **Note 15**) to generate visible PCR product. If the primary PCR requires more than 30 cycles, the problems described in the previous two items will usually appear.

If you want to perform MOS, please follow the following procedure for PCR amplification using the second hybridization solution (**Subheading 3.1.5.2., step 9**).

3.2.1. PCR Amplification for MOS

If the complexity of tester and driver samples is very high or if the difference in gene expression between tester and driver is very small, one can plan to perform MOS from the beginning of the experiment. In that case, after subtractive hybridization (**Subheading 3.1.5.**), perform PCR amplification using the following protocol instead of using protocol in **Subheading 3.1.6.** If you have already made the SSH subtracted library and found high background upon differential screening, you have the option to perform MOS on the SSH-generated library. You can use the hybridization mix generated in **Subheading 3.1.5.2., step 9**) for PCR amplification using the following protocol.

3.2.1.1. PRIMARY PCR-1

1. Transfer 10 μL of each diluted second hybridization (from **Subheading 3.1.5.**) into appropriately labeled tubes (*see* **Note 12**).
2. Prepare a Master Mix for the primary PCR-1. For each reaction, combine the reagents as follows.

Component	Amount per reaction
Sterile H_2O	92.5 μL
10X PCR buffer	12.5 μL
dNTP mixture (10 m*M* each)	2.5 μL
PCR Primer 1	5.0 μL
50X polymerase mixture	2.5 μL
Total volume	115 μL

3. Mix well and briefly centrifuge the tube.
4. Place 115 μL aliquot of Master Mix into each reaction tube from **step 1**.
5. Place 125 μL aliquot of final mix into five 0.5 μL PCR tubes (25 μL per tube).
6. Overlay with one drop of mineral oil.
7. Incubate the reaction mixture in a thermal cycler at 72°C for 5 min to extend the adaptors (*see* **Note 13**).
8. Immediately commence thermal cycling (*see* **Note 17** to calculate the number of PCR cycles you need):

   95°C   10 s
   66°C   10 s
   72°C   1.5 min

9. Combine 2 μL of each (of 5) primary PCR-1 product in one tube and add 390 μL of H_2O.
10. Place 1 μL aliquot of each diluted primary PCR-1 product mixture from **step 9** into an appropriately labeled PCR tube.
11. Prepare Master Mix for primary PCR-2 as follows.

Component	Amount per reaction
Sterile H_2O	19.5 μL
10X PCR buffer	2.5 μL
dNTP mixture (10 m*M* each)	0.5 μL
PCR Primer 1	1.0 μL
50X polymerase mixture	0.5 μL
Total volume	24 μL

12. Mix well and briefly centrifuge the tube.
13. Place 24 μL aliquot of Master Mix into each reaction tube from **step 10**.
14. Overlay with one drop of mineral oil.
15. Immediately commence thermal cycling:

   10 cycles:
   94°C   30 s
   66°C   30 s
   72°C   1.5 min

16. Analyze 4 μL from each reaction on a 2% agarose/EtBr gel.

3.2.1.2. SECONDARY PCR

1. Dilute 2 μL of each primary PCR-2 product generated in **Subheading 3.2.1.1., step 16** in 38 μL of H_2O.
2. Place 2 μL aliquot of each diluted primary PCR-2 product into an appropriately labeled tube.
3. Prepare a Master Mix for secondary PCR. For each reaction, combine the reagents as follows:

Component	Amount per reaction
Sterile H_2O	37.0 μL
10X PCR buffer	5.0 μL
dNTP mixture (10 mM each)	1.0 μL
PCR Primer NP1	2.0 μL
PCR Primer NP2R	2.0 μL
50X Polymerase mixture	1.0 μL
Total volume	48.0 μL

4. Mix well and briefly centrifuge the tube.
5. Place 48 μL aliquot of Master Mix into each reaction tube from **step 2**.
6. Overlay with one drop of mineral oil.
7. Immediately commence thermal cycling:
   10 cycles:
   95°C   10 s
   68°C   10 s
   72°C   1.5 min
8. Analyze 4 μL from each tube on a 2% agarose/EtBr gel.
9. The PCR product of secondary PCR is purified by phenol/chloroform extraction and ethanol precipitation (*see* **Note 4**).
10. Dissolve the pellet in 20–40 μL of NT buffer up to concentration 20 ng/μL of DNA.
11. Analyze 2 μL of purified PCR product from **step 9** on a 2% agarose/EtBr gel.
12. Dilute 1 μL of purified PCR product from **step 9** in 1.6 mL H_2O (this will be your undigested control).
13. Store at −20°C

3.2.2. Xma*I Digestion*

1. Add the following reagents into the tube.
   12 μL H_2O
   2 μL 10X *Xma* I restriction buffer
   5 μL DNA (**Subheading 3.2.1.2., step 10**)
   1 μL *Xma* I (10 U/μL)
2. Mix and incubate at 37°C for 2 h.
3. Add 2 μL of 0.2 *M* EDTA to terminate the reaction.

4. Incubate at 70°C for 5 min to inactivate enzyme.
5. Store at −20°C.

3.2.3. MOS Hybridization

Combine the following reagents in a fresh 1.5-mL tube:

2 μL H_2O2 μL
1 μL *Xma* I digested DNA
1 μL 4X Hybridization buffer

Place 2 μL of this mixture in a 0.5 mL microcentrifuge tube and overlay with one drop of mineral oil.

Incubate in a thermal cycler at 98°C for 1.5 min.

Incubate in a thermal cycler at 68°C for 3 h.

Add 200 μL of dilution buffer to the tube and mix by pipeting.

Heat in a thermal cycler at 70°C for 7 min.

Store at −20°C

3.2.4. MOS PCR Amplification

1. Prepare a Master Mix for all MOS PCR reactions as follows:

Component	Amount per reaction
Sterile H_2O	19.5 μL
10X PCR buffer	2.5 μL
dNTP mixture (10 m*M* each)	0.5 μL
MOS PCR Primer (NP2Rs)	1.0 μL
50X polymerase mixture	0.5 μL
Total volume	24.0 μL

2. Add 1 μL of each diluted cDNA sample (after hybridization and the corresponding undigested control) to an appropriately labeled tube containing 24 μL of Master Mix.
3. Overlay with one drop of mineral oil.
4. Incubate the reaction mix in a thermal cycler at 72°C for 2 min to extend the adaptors. (Do not remove the samples from the thermal cycler.)
5. Immediately commence thermal cycling:

19 cycles:
94°C   30 s
62°C   30 s
72°C   1.5 min

6. Analyze 4 μL from each tube on 2% agarose/EtBr gel.

3.3. Cloning of Subtracted cDNAs

Once a subtracted sample is confirmed to be enriched in cDNAs derived from differentially expressed genes, the PCR products (from **Subheading 3.1.6.**, secondary PCR or from **Subheading 3.2.4.**, MOS PCR amplification) can be subcloned using several conventional cloning techniques. The following describes two such methods that are currently used.

1. T/A cloning

    Use 3 μL of the secondary PCR product (**Subheading 3.1.5.4.**, **step 8**) or MOS PCR product (**Subheading 3.2.4.**, **step 6**) for cloning with a T/A-based system, such as the AdvanTAge PCR Cloning Kit (Invitrogen), according to the manufacturer's protocol. The library is transformed into bacteria (electrocompetent cells) by electroporation (1.8 kV) using a pulser (BioRad) and plated onto agar plates containing ampicillin, X-Gal, and IPTG. Recombinant (white colonies) clones are picked and used to inoculate LB medium in 96-well microtiter plates. Bacteria should be allowed to grow at 37°C for 4 h before insert amplification (**Subheading 3.4.2.**). Typically, 10^4 independent clones from 1 μL of secondary PCR product can be obtained using the above cloning system and electroporation. It is important to optimize the cloning efficiency because a low cloning efficiency will result in a high background.

2. Site-specific or blunt-end cloning

    For site-specific cloning, cleave at the *Eag* I, *Not* I, and *Xma* (*Sma* I, *Srf* I) sites embedded in the adaptor sequences and then ligate the products into an appropriate plasmid vector. Keep in mind that all of these sites might be present in the cDNA fragments. The *Rsa* I site in the adaptor sequences can also be used for blunt-ended cloning. Commercially available cloning kits are suitable for these purposes.

    The number of independent colonies obtained for each library depends on the estimated number of differentially expressed genes, as well as the subtraction and subcloning efficiencies. In general, 500 colonies can be initially arrayed and studied. The complexity of the library can be increased by additional subcloning of secondary PCR products (from **Subheading 3.1.6.**) or MOS PCR products (from **Subheading 3.2.4.**).

3.4. Differential Screening of the Subtracted cDNA Library

Two approaches can be utilized for differential screening of the arrayed subtracted cDNA clones; cDNA dot blots and colony dot blots. For colony dot blots, bacterial colonies are spotted on nylon filters, grown on antibiotic plates, and processed for colony hybridization. This method is cheaper and more convenient, but it is less sensitive and gives a higher background than PCR-based cDNA dot blots. The cDNA array approach is highly recommended (**Subheading 3.4.2.**).

3.4.1. Amplification of cDNA Inserts by PCR

For high-throughput screening, a 96-well format PCR from one of several thermal cycler manufacturers is recommended. Alternatively, single tubes can be used.

1. Randomly pick 96 white bacterial colonies.
2. Grow each colony in 100 μL of LB-amp medium in a 96-well plate at 37°C for at least 2 h (up to overnight) with gentle shaking.
3. Prepare a master mix for 100 PCR reactions (*see* **Note 18**):

Reagent	Amount per reaction
10X PCR reaction buffer	2.0 μL
Nested Primer NP1*	0.6 μL
Nested Primer NP2R*	0.6 μL
dNTP Mix (10 mM)	0.4 μL
H$_2$O	15.0 μL
50X Advantage cDNA PCR Mix	0.4 μL
Total volume	19.0 μL

Alternatively, primers flanking the insertion site of the vector can be used in PCR amplification of the inserts.

4. Place 19 μL aliquot of the master mix into each tube or well of the reaction plate.
5. Transfer 1 μL of each bacterial culture (from **step 2**) to each tube or well containing master mix (*see* **Note 19**).
6. Perform PCR in an oil-free thermal cycler with the following conditions:
   1 cycle:
   94°C   2 min
   then 22 cycles:
   94°C   30 s
   68°C   3 min
7. Analyze 5 μL from each reaction on a 2% agarose/EtBr gel (*see* **Note 20**).

3.4.2. Preparation of cDNA Dot Blots of the PCR Products

1. For each PCR reaction, combine 5 μL of the PCR product and 5 μL of 0.6 M NaOH (freshly made or at least freshly diluted from concentrated stock).
2. Transfer 1–2 μL of each mixture to a nylon membrane. This can be accomplished by dipping a 96-well replicator in the corresponding wells of a microtiter dish used in the PCR amplification and spotting it onto a dry nylon filter. Make at least two identical blots for hybridization with subtracted and reverse-subtracted probes (*see* **Subheading 3.1.3.**) (*see* **Note 22**).
3. Neutralize the blots for 2–4 min in 0.5 M Tris-HCl (pH 7.5) and wash in 2X SSC.
4. Immobilize cDNA on the membrane using a UV crosslinking device (such as Stratagene's UV Stratalinker), or bake the blots for 4 h at 68°C.

3.4.3. Differential Hybridization With Tester and Driver cDNA Probes

Label tester and driver cDNA probes by random-primer labeling using a commercially available kit. The hybridization conditions given here are optimized for BD Biosciences Clontech's ExpressHyb solution; the optimal hybridization conditions for other systems should be determined empirically.

The following four different probes will be used for differential screening hybridization:

(1) Tester-specific subtracted probe (forward-subtracted probe)
(2) Driver-specific subtracted probe (reverse-subtracted probe)
(3) cDNA probe synthesized directly from tester mRNA

(4) cDNA probe synthesized directly from driver mRNA
(5) *see* **Note 23**.

1. Prepare a prehybridization solution for each membrane:
   a. Combine 50 μL of 20X SSC, 50 μL of sheared salmon sperm DNA (10 mg/ mL), and 10 μL of blocking solution (containing 2 mg/mL of unpurified NP1, NP2R, cDNA synthesis primers and their complementary oligonucleotides).
   b. Boil the blocking solution for 5 min, then chill on ice.
   c. Combine the blocking solution with 5 mL of ExpressHyb Hybridization Solution (BD Biosciences Clontech).
2. Place each membrane in the prehybridization solution prepared in **step 1**. Prehybridize for 40–60 min with continuous agitation at 72°C.
   Note: It is important to add blocking solution in prehybridization solution. Because subtracted probes contain the same adaptor sequences as arrayed clones, these probes hybridize to all arrayed clones regardless of the sequences.
3. Prepare hybridization probes:
   a. Mix 50 μL of 20X SSC, 50 μL of sheared salmon sperm DNA (10 mg/mL) and 10 μL blocking solution, and purified probe (at least 10^7 cpm per 100 ng of subtracted cDNA). Make sure the specific activity of each probe is approximately equal.
   b. Boil the probe for 5 min, then chill on ice.
   c. Add the probe to the prehybridization solution.
4. Hybridize overnight with continuous agitation at 72°C.
5. Prepare low-stringency (2X SSC/0.5% SDS) and high-stringency (0.2X SSC/0.5% SDS) washing buffers and warm them up to 68°C.
6. Wash membranes with low-stringency buffer (4 X 20 min at 68°C), then wash with high-stringency buffer (2 X 20 min at 68°C).
7. Perform autoradiography.
8. If desired, remove probes from the membranes by boiling for 7 min in 0.5% SDS. Blots can typically be reused at least five times.
   Note: To minimize hybridization background, store the membranes at −20°C when they are not in use.

4. Notes

1. Self-subtraction (with both Tester and Driver prepared from one DNA sample) is recommended as the best comprehensive control. In the self-subtracted control, subtracted secondary PCR requires more cycles than unsubtracted secondary PCR. A number of other control experiments may be performed for fast analysis of SSH and MOS experiments (**Table 2**).
2. For experimental systems such as transfection, overexpression, mRNA injection, or viral infection using mammalian or viral expression systems, we strongly recommend that you use affecting RNA/DNA sequence for compensation of overexpressed sequence concentration.
   For example, if you are searching for p53-up-regulated genes in a p53 overexpressed cell line, add *Rsa* I-digested p53 cDNA into *Rsa* I-digested driver sample

Table 2
List of Control Experiments That Can Be Performed With SSH and MOS

Control	Controlled factor	Expected results
Unsubtracted control (SSH)	Subtraction efficiency	Differences between control and PCR2 gel-pattern
Unhybridized control (SSH)	Hybridization process	Control PCR2 need more cycles
Undigested control (MOS)	Subtraction efficiency, MOS efficiency	Differences between control and MOS-PCR gel-pattern

(about 1/10 of driver cDNA concentration) after you prepare adaptor-ligated tester. Adding exogenous DNA/RNA earlier (in RNA sample) or before *Rsa* I digestion may cause disproportion of this material in initial DNAs.

3. We recommend the use of poly(A)+ RNA as starting material. Amplified cDNA should be used as a starting material only when enough RNA is not available. The amplification of two cDNA samples to be subtracted is a crucial procedure and any disproportion during cDNA amplification may cause artifacts in the subtraction results. Some RNA types cannot be amplified because the messages are too long and are not available for subtraction and analysis.

4. Phenol-chloroform extraction and ethanol precipitation:
   a. Add equal volumes of phenol:chloroform:isoamyl alcohol (25:24:1) and vortex thoroughly.
   b. Centrifuge the tubes at 14,000g for 10 min.
   c. Remove the top aqueous layer and transfer to a fresh microcentrifuge tube.
   d. Add equal volumes of chloroform:isoamyl alcohol (24/1) and vortex thoroughly.
   e. Centrifuge the tubes at 14,000g for 10 min.
   f. Remove the top aqueous layer and transfer to a fresh microcentrifuge tube.
   g. Add 0.5 volume of 4M NH4OAc, mix, then add 2.5 volumes of 95% ethanol and vortex thoroughly.
   h. Centrifuge the tubes at 14,000g for 20 min.
   i. Remove the supernatant carefully.
   j. Add 200 µL of 80% ethanol.
   k. Centrifuge the tubes at 14,000g for 10 min.
   l. Remove the supernatant carefully.
   m. Air dry the pellets for 5–10 min.
   n. Dissolve the pellets in appropriate volume of TN buffer.

5. Using glycogen or any type of coprecipitants during DNA precipitation may increase viscosity of DNA solution and prevent DNA hybridization in some cases. We recommend avoiding use of these reagents if possible.

6. We do not recommend using silica matrix-based PCR purification systems at this stage.

7. Water may denature short DNA fragments and may make the adaptor ligation difficult. We advise using TN buffer.

9. Ligation efficiency test:

   a. Place 1 μL aliquot of each undiluted unsubtracted control sample (**Subheading 3.1.4.**, **step 8**) into an appropriately labeled 0.5-ml PCR tube.

   b. Prepare a Master Mix for all of the reaction tubes. Combine the reagents as follows:

Component	Amount per reaction
Sterile H$_2$O	19.5 μL
10X PCR buffer	2.5 μL
dNTP mixture (10 mM each)	0.5 μL
PCR Primer P1	1.0 μL
50X polymerase mixture	0.5 μL
Total volume	24.0 μL

   c. Mix and briefly centrifuge the tubes.

   d. Place 24 μL aliquot of Master Mix into each of the reaction tubes prepared in **step 1**.

   c. Overlay with one drop of mineral oil.

   d. Incubate the reaction mixture in a thermal cycler at 72°C for 5 min to extend the adaptors.

   e. Immediately commence thermal cycling:

   > 15 cycles:
   > 95°C   10 s
   > 66°C   10 s
   > 72°C   1.5 min

   f. Analyze 4 μL from each tube on a 2% agarose/EtBr gel. This PCR product should have a similar pattern to that of *Rsa* I-digested DNA. If PCR products are not visible after 15 cycles, perform three more cycles and again analyze the PCR product. If PCR products are not visible after 21 cycles, the activity of the polymerase mixture needs to be examined. If there is no problem with the polymerase mixture, the ligation reaction should be repeated with fresh samples before proceeding to the next step.

10. Recommended first hybridization times for different DNA samples:

    Sample type

    First hybridization time

    Bacterial genome subtraction

    1–3 h

    Eukaryotic genome subtraction

    3–5 h

    cDNA subtraction

    7–12 h

10. We recommend that you use two blocks thermal cycler (or two thermal cyclers nearby) for proper and fast operations.

11. We recommend transferring this tube immediately after denaturing (98°C for 1.5 min) into thermal cycler with first hybridization process (68°C) and waiting for 1 min before proceeding to the next step.

12. If hybridization mix was frozen, we recommend the following before proceeding with PCR reactions: mix hybridization samples well by pipeting, heat in a thermal cycler at 72°C for 7 min, then mix again by pipeting and use for PCR.

13. This step "fills in" the missing strand of the adaptors and thus creates binding sites for the PCR primers.

14. For some complicated subtractions (with complex tissues or eukaryotic genomes), we recommend performing primary PCR two times one by one. This procedure may significantly reduce background, generated by partial disruption of PCR-suppression effect. First, perform primary PCR as described in **Subheading 3.1.5.3.** Then perform another primary PCR as follows:

    a. Dilute 2 µL of each primary PCR product (from **step 7**) in 78 µL of H_2O.
    b. Place 1 µL of each diluted primary PCR product from step 1 into appropriately labeled tube.
    c. Combine the following reagents to prepare a Master Mix for each reaction.

Component	Amount per reaction
Sterile H_2O	19.5 µL
10X PCR buffer	2.5 µL
dNTP mixture (10 mM each)	0.5 µL
PCR Primer P1	1.0 µL
Advantage cDNA PCR mix	0.5 µL
Total volume	24.0 µL

    d. Mix well and briefly centrifuge the tube.
    e. Place 24 µL aliquot of Master Mix into each reaction tube from **step 2**.
    f. Overlay with one drop of mineral oil.
    g. Immediately commence thermal cycling:
       10 cycles:
       95°C   10 s
       66°C   10 s
       72°C   1.5 min
    h. Analyze 4 µL from each tube on a 2% agarose/EtBr gel, then proceed to secondary PCR (**Subheading 3.1.6.**).

15. If the SSH primary PCR requires more then 36 cycles, "in vitro cloning" will occur. As a result, only false-positive clones may be found during differential screening procedure.

16. To illustrate the utility of combining SSH and MOS for eukaryotic genome comparison, we will describe our efforts to isolate genes that are present in one freshwater planaria strain but are absent in another. In this study, we used two closely related strains of freshwater planaria *Girardia tigrina* that reproduce in different

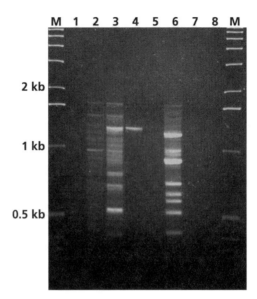

Fig. 4. **Lane 1**: MOS PCR product of undigested control of BB self-subtraction.
Lane 2: MOS PCR product of BB self-subtraction. **Lane 3**: MOS PCR product of BA
experimental subtraction. **Lane 4**: MOS PCR product of undigested control of BA experimental subtraction. **Lane 5**: MOS PCR product of undigested control of AB experimental subtraction. **Lane 6**: MOS PCR product of AB experimental subtraction. **Lane 7**:
MOS PCR product of AA self-subtraction. **Lane 8**: MOS PCR product of undigested
control of AA self-subtraction.

ways. Whereas one strain has exclusively asexual reproduction, the other repro-
duces both sexually and asexually. We compared the genomes of both strains of
G. tigrina to search for genetic determinants of asexuality.

Total DNA from these strains was purified using the procedure described in
Subheading 3.1.1. *(12)*. The SSH and MOS combination was used to isolate genes
that are differentially present in each planaria strain. Forward subtraction (AB)
was performed using asexual DNA (sample A) as tester and sexual DNA (sample
B) as driver, and the forward-subtracted DNA was enriched for DNA fragments
specific to the asexual strain of freshwater planaria. Reverse-subtracted DNA (BA)
was enriched for DNA fragments specific to the sexual planaria strain. Self-subtractions were performed for both DNA samples to get a quick idea of subtraction
efficiency. Subsequent MOS-PCR analysis confirmed that the self-subtractions (as
well as undigested controls) require more PCR cycles to generate visible PCR
product, indicating that the subtraction was successful (**Fig. 4**).

We anticipated that the differences between tester and driver DNA would be
small, so we proceeded with a differential screening procedure (**Fig. 5**). Eighty-
six randomly selected clones from each (forward- and reverse-subtracted) library

Forward subtraction (A-B) library differential screening

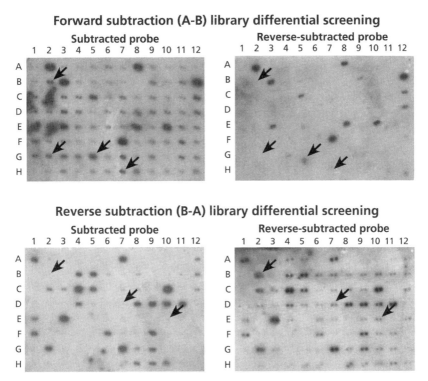

Fig. 5. Differential screening approach. Top panel represents clones from a forward-subtracted library (AB) and bottom panel represents clones from a reverse-subtracted library (BA). Two identical dot blots are prepared from each subtracted library. Dot blots from both the libraries are hybridized with DNA probes made from forward-subtracted DNA (AB) and reverse-subtracted DNA (BA).

were arrayed (DNA dot blot) onto nylon membranes. DNA dot blots were hybridized to probes prepared from the subtracted and reverse-subtracted libraries. **Figure 5** shows typical results of differential screening of a subtracted DNA libraries obtained using the SSH and MOS combination. These results reveal the following types of clones:

a. Clones hybridizing to the one probe only. These clones correspond to the differentially presented DNA, but must be verified by Southern blot analysis. The signal intensity depends on the copy number in genomic or extrachromosomal DNA.

b. Clones hybridizing to both subtracted probes with the same efficiency. These clones do not correspond to the differentially presented DNA; this is background.

c. Clones hybridizing to both subtracted probes with different hybridization efficiencies. In the case of genomic DNA subtraction, these clones may represent

Fig. 6. PCR analysis of differential genes found via suppression subtractive hybridization and mirror orientation selection (MOS) of two freshwater planaria genomes. S-polymerase chain reaction (PCR) from sexual planaria genomic DNA; A-PCR from asexual planaria genomic DNA. pAS14- and pAS22-clones from MOS library specific for asexual planaria. pSA32- and pSA34-clones from MOS library specific for sexual planaria. Control is a housekeeping gene present in both genomes.

genes (DNA fragments) with different number of copies per genome. In the case of cDNA subtraction, these clones do represent differentially expressed clones. In some cDNA subtractions, this difference can be a result of random fluctuation and does not represent differentially expressed cDNA. For this reason, it is always recommended to confirm true differential expression of these clones by Northern analysis or RT-PCR.

    d. Clones that do not hybridize noticeably to either hybridization probe. These clones may not contain DNA insertion or may be present at very low concentration in the subtracted probe. (In most cases, they do not represent differentially presented clones.)

Differential screening revealed about 60% and 30% of the strain-specific clones in AB and BA libraries, respectively. About 50% of the asexual-specific clones turned out to be a novel extrachromosomal DNA-containing virus-like element. Several strain-specific genes were identified as lectins. We randomly selected two clones from each library for confirmation of differential expression by PCR (**Fig. 6**). Most of the nondifferential DNA sequences were identified as the mariner element, approx 7000 copies that were present in each compared genome *(18)*.

17. The recommended number of primary PCR1 cycles for MOS is the number of SSH primary PCR minus 2. For example, if the SSH primary PCR was visible on agarose/EtBr gel after 31 cycles, you will need $31 - 2 = 29$ cycles of primary PCR1 for MOS.

18. Short PCR primers NP1s and NP2s can be used for insert amplification to reduce hybridization background. However, this is not always necessary.

19. Freshly grown 96-well plates should be used for PCR before bacterial cells precipitate, otherwise 1-μL aliquots will not be equal.

20. It is possible that approx 5–10% of clones will not yield PCR product as a result of imperfect cloning.

21. The protocol uses 15 ng of ligated tester cDNA and 450 ng of driver cDNA. The ratio of driver to tester can be changed if different subtraction efficiency is desired.

22. We highly recommend that you make four identical blots. Two of the blots will be hybridized to forward and reverse subtracted cDNAs and the other two can be hybridized to cDNA probes synthesized from tester and driver mRNAs.

23. The first two probes are the secondary PCR products (**Subheading 3.1.5.4., step 8 or 3.2.1.2., step 10**) of the subtracted cDNA pool. The last two cDNA probes can be synthesized from the tester and driver poly(A)+ RNA. They can be used as either single-stranded or double-stranded cDNA probes (**Subheading 3.1.2.1. and 3.2.1.2.**). Alternatively, unsubtracted tester and driver cDNA (**Subheading 3.1.5.4., step 8 or 3.2.1.2., step 10**) or preamplified cDNA from total RNA (*11*) can be used if enough poly(A)+ RNA is not available. If you have made the MOS-subtracted library, you can still screen it using the same probes.

Acknowledgments

We thank Dr. L. Diatchenko and S. Trelogan for critical reading of the manuscript and Anna Sayre for preparing the figures for this chapter.

References

1. Luk'ianov, S. A., Gurskaya, N. G., Luk'ianov, K. A., Tarabykin, V. S., and Sverdlov, E. D. (1994) Highly efficient subtractive hybridization of cDNA. *J. Bioorgan. Chem.* **20**, 386–388.

2. Gurskaya, N. G., Diatchenko, L., Chenchik, A., et al. (1996) Equalizing cDNA subtraction based on selective suppression of polymerase chain reaction: cloning of Jurkat cell transcripts induced by phytohemaglutinin and phorbol 12-myristate 13-acetate. *Anal. Biochem.* **240**, 90–97.

3. Akopyants, N. S., Fradkov, A., Diatchenko, L., et al. (1998) PCR-based subtractive hybridization and differences in gene content among strains of *Helicobacter pylori*. *Proc. Natl. Acad. Sci. USA* **95**, 13108–13113.

4. Diatchenko, L., Lau, Y. F. C., Campbell, A. P., et al. (1996) Suppression subtractive hybridization: a method for generating differentially regulated or tissue-specific cDNA probes and libraries. *Proc. Natl. Acad. Sci. USA* **93**, 6025–6030.

5. Lukyanov, K. A., Launer, G. A., Tarabykin, V. S., Zaraisky, A. G., and Lukyanov, S. A. (1995) Inverted terminal repeats permit the average length of amplified DNA fragments to be regulated during preparation of cDNA libraries by polymerase chain reaction. *Anal. Biochem.* **229**, 198–202.

6. Siebert, P. D., Chenchik, A., Kellogg, D. E., Lukyanov, K. A., and Lukyanov, S. A. (1995) An improved PCR method for walking in uncloned genomic DNA. *Nucleic Acids Res.* **23**, 1087–1088.

7. Jin, H., Cheng, X., Diatchenko, L., Siebert, P. D., and Huang, C. C. (1997) Differential screening of a subtracted cDNA library: a method to search for genes preferentially expressed in multiple tissues. *BioTechniques* **23**, 1084–1086.

8. Desai, S., Hill, J., Trelogan, S., Diatchenko, L., and Siebert, P. (2000) Identification of differentially expressed genes by suppression subtractive hybridization, in *Functional Genomics* (Hunt, S. P. and Livesey, F. J., eds.), Oxford University Press, 81–111.

9. Rebrikov, D. V., Britanova, O. V., Gurskaya, N. G., Lukyanov, K. A., Tarabykin, V. S., and Lukyanov, S. A. (2000) Mirror orientation selection (MOS): a method for eliminating false positive clones from libraries generated by suppression subtractive hybridization. *Nucleic Acids Res.* **28,** e90.

10. Gubler, U. and Hoffman, B. J. (1983) A simple and very efficient method for generating cDNA libraries. *Gene* **25,** 263–269.

11. Chenchik, A., Zhu, Y. Y., Diatchenko, L., Li, R., Hill, J., and Siebert, P. D. (1998) Generation and Use of High-Quality cDNA from Small Amounts of Total RNA by SMART PCR in *Gene Cloning and Analysis by RT-PCR* (Siebert, P. D. and Larrick, J. W., eds.), Molecular Laboratory Methods Number 1, 305–319.

12. Britten, R. J. and Davidson, E. H. (1985) In *Nucleic Acid Hybridization- A Practical Approach* (Hames, B. D. and Higgins, S., eds.), IRL Press, Oxford, 3–15.

13. Kellogg, D. E., Rybalkin, I., Chen, S., et al. (1994) TaqStart Antibody: "hot start" PCR facilitated by a neutralizing monoclonal antibody directed against Taq DNA polymerase. *BioTechniques* **16,** 1134–1137.

14. Sambrook, J., Fritsch, E. F., and Maniatis, T. (1989) *Molecular Cloning, A Laboratory Manual.* Cold Spring Harbor Lab., Cold Spring Harbor.

15. Ausubel, F. M., Brent, R., Kingston, R. E., et al. (1994) *Current Protocols in Molecular Biology.* Greene Publishing Associates and John Wiley & Sons, Inc., NY. **1,** Ch. 2.4.

16. Chomczynski, P. and Sacchi, N. (1987) Single-step method of RNA isolation by acid guanidinium thiocyanate-phenol-chlorophorm extraction. *Anal. Biochem.* **162,** 156–159.

17. Farrell, R. E., Jr. (1993) *RNA Methodologies: A Guide for Isolation and Characterization.* Academic, San Diego, CA.

18. Garcia-Fernandez, J., Marfany, G., Baguna, J., and Salo, E. (1993) Infiltration of mariner elements. *Nature* **364,** 109–110.

9

Small Amplified RNA-SAGE

Catheline Vilain and Gilbert Vassart

Summary

Serial analysis of gene expression (SAGE) is a powerful genome-wide analytic tool to determine expression profiles. Since its description in 1995 by Victor Velculescu et al., SAGE has been widely used. Recently, the efficiency of the method has been emphasized as a means to identify novel transcripts or genes that are difficult to identify by conventional methods. SAGE is based on the principle that a 10-base pair (bp) cDNA fragment contains sufficient information to unambiguously identify a transcript, provided it is isolated from a defined position within this transcript. Concatenation of these sequence tags allows serial analysis of transcripts by sequencing multiple tags within a single clone. Extraction of sequence data by computer programs provides a list of sequence tags that reflect both qualitatively and quantitatively the gene expression profile. Several modifications to the initial protocol allowed to start from 1 µg total RNA (or 10^5 cells). In order to reduce the amount of input RNA, protocols including extra polymerase chain reaction (PCR) steps were designed. Linear amplification of the mRNA targets might have advantage over PCR by minimizing biases introduced by the amplification step; therefore we devised a SAGE protocol in which a loop of linear amplification of RNA has been included. Our approach, named "small amplified RNA-SAGE" (SAR-SAGE) included a T7 RNA polymerase promoter within an adapter derived from the standard SAGE linker. This allowed transcription of cDNA segments, extending from the last *Nla*III site of transcripts to the polyA tail; these small amplified RNAs then serve as template in a classical (micro)SAGE procedure. As the cDNAs are immobilized on oligo(dT) magnetic beads, several rounds of transcription can be performed in succession with the same cDNA preparation, with the potential to increase further the yield in a linear way. Except for the transcription step itself, the present procedure does not introduce any extra enzymatic reaction in the classical SAGE protocol, it is expected to keep the representation biases associated with amplification as low as possible.

Key Words: PCR, SAGE, sequencing, serial analysis of gene expression, linear amplification

From: *Methods in Molecular Biology, Vol. 258: Gene Expression Profiling: Methods and Protocols*
Edited by: R. A. Shimkets © Humana Press Inc., Totowa, NJ

1. Introduction

Serial analysis of gene expression (SAGE) *(1)* is a powerful genome-wide analytic tool to determine expression profiles. Since its description in 1995 by Victor Velculescu et al. *(1)*, SAGE has been widely used in fields as diverse as cancer research, development, and study of microorganisms *(2–4)*. The efficiency of the SAGE method has been emphasized recently as a means to identify novel transcripts or genes that are difficult to identify by conventional methods *(5)*. SAGE *(1)* is based on the principle that a 10-base pair (bp) cDNA fragment contains sufficient information to unambiguously identify a transcript, provided it is isolated from a defined position within this transcript. Linking of sequence tags allows serial analysis of transcripts by sequencing multiple tags within a single clone. Extraction of sequence data by computer programs provides a list of sequence tags that reflects the gene expression profile, qualitatively and quantitatively. Publicly available, SAGE libraries can easily be compared. The SAGE procedure is described in **Fig. 1**.

The initial SAGE protocol was not suitable for the study of small samples, as it required 5 µg of polyA RNA, approximately 5×10^6 cells of input RNA.

The first improvements to reduce the amount of starting RNA were brought by the SAGE Adaptation for Downsized Extract (SADE) procedure *(6)*. The researchers optimized steps of the protocol, diminishing the loss of material, and increasing robustness of the method. These modifications generated libraries from as few as 30,000 cells. Some modifications were also brought to the classical SAGE protocol and the resulting protocol is known as microSAGE *(7)* (www.sagenet.org/sage-protocol.htm). A MiniSAGE *(8)* procedure has been described, that allows starting RNA amounts from 1 µg total RNA.

For some research purposes, 5×10^4 cells or 1 µg total RNA is an unachievable prerequisite. This may be true when subgroups of cells have to be isolated, for instance by laser microdissection. To apply the SAGE technology to small samples a prior amplification step is required. Different protocols have been designed for this purpose *(9–12)*, and they all rely on polymerase chain

---

Fig. 1. (*Opposite page*) Classical SAGE procedure. RNA is bound to oligo (dT) magnetic beads for cDNA synthesis. cDNA is cut with a restriction endonuclease named "anchoring enzyme," (a restriction endonuclease with 4-bp recognition site was chosen in order to cleave theoretically every transcript at least once). The cDNAs are divided in half and ligated to linkers containing a type IIS restriction site. The type IIS restriction endonuclease, named "tagging enzyme," cleaves at a defined distance 14 bp away from its asymmetric recognition site; this releases linkers with a short piece of the cDNA constituting a 10 bp sequence "tag." After blunt ending, the two pools of tags are ligated together to form ditags. Ditags are amplified by polymerase chain reaction

SAGE PROCEDURE

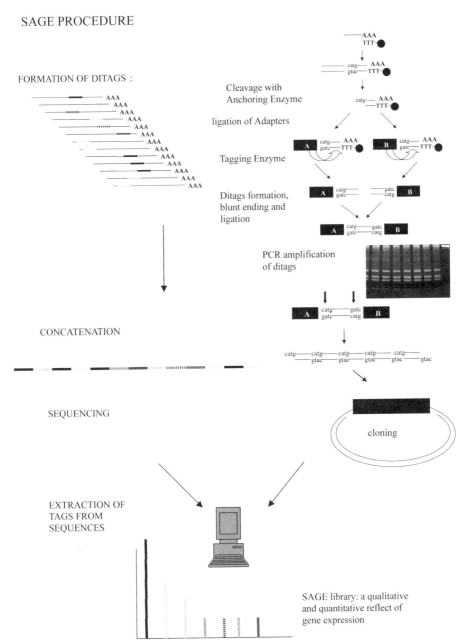

(PCR). After cleavage of the amplified ditags with anchoring enzyme, 26 bp ditags are isolated from the linkers and ligated together to form concatemers. The concatemers, which consist of ditags separated by a 4-bp punctuation corresponding to the anchoring enzyme restriction site, are cloned and sequenced.

reaction (PCR) amplification. PCR-SAGE *(11)* and SAGE-Lite *(9)* both rely on the inherent poly(C) terminal transferase activity of reverse transcriptase to switch templates during DNA polymerization (Clontech SMART system). Starting with nine human GV oocytes (900 pg polyA RNA) PCR-SAGE *(11)* generated a library validated by the presence of transcripts known to be expressed in oocytes. SAGE-Lite *(9)*, starting with 50 ng of total RNA derived from cerebrovascular tissue or HT1080 cells, was validated in a similar way. In the protocol proposed by Lee et al. *(12)* the PCR amplification is performed following ligation of the SAGE linkers and product purification before subsequent use in the SAGE procedure. The MicroSAGE technique developed by Datson et al. *(10)* included extra PCR amplification of the ditags, but increased the amount of duplicate dimers in final result.

All these methods include additional PCR steps, either before or after the generation of the SAGE tags. Linear amplification of the mRNA targets might have advantage over PCR by minimizing biases introduced by the amplification step. We devised a SAGE protocol *(13)* in which a loop of linear amplification of RNA *(14)* has been included. Our approach, small amplified RNA-SAGE (SAR-SAGE) *(13)* included a T7 RNA polymerase promoter within an adapter derived from the standard SAGE linker (**Fig. 2**). This allowed transcription of cDNA segments, extending from the last *Nla*III site of transcripts to the polyA tail; these small amplified RNAs then serve as template in a classical (micro)SAGE procedure *(7)*. The cDNAs are immobilized on oligo(dT) magnetic beads, therefore several rounds of transcription can be performed in succession with the same cDNA preparation, with the potential to further increase the yield in a linear way *(15)*. Except for the transcription step itself, the present procedure does not introduce extra enzymatic reaction in the classical SAGE protocol, it is expected to keep the representation biases associated with amplification as low as possible. Indeed, we observed an overall correlation of 75% when we compared a SAR-SAGE library and a classical microSAGE library *(7)* prepared from 50 ng and 5 μg of the same total RNA preparation, respectively. In the following sections, we describe the procedure to generate small-amplified RNA: binding of mRNA to magnetic beads, cDNA synthesis, cleavage with anchoring enzyme, ligation of T7-SAGE adapter containing the promoter of the T7-RNA polymerase, and in vitro transcription. Small amplified RNA serve then as template in the classical SAGE procedure.

2. Materials

To avoid the loss of beads on microcentrifuge tube walls, use sterile, siliconized microcentrifuge tubes (Ambion No-stick Rnase Free 1.5 mL microcentrifuge tubes cat. no. 12450 are often recommended).

SAR-SAGE procedure

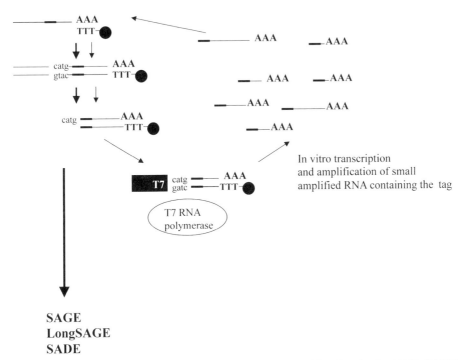

In vitro transcription
and amplification of small
amplified RNA containing the tag

SAGE
LongSAGE
SADE

Fig. 2. Scheme of the additional loop of mRNA amplification in the SAR-SAGE procedure. Ligation of an adapter derived from the standard SAGE linker, containing the T7 RNA polymerase promoter allows transcription of cDNA segments, extending from the last *Nla*III site of transcripts to the polyA tail.

2.1. mRNA Binding on Oligo(dT) Magnetic Beads

1. Magnetic stand.
   The following items must be RNase-free:
2. Dynabeads oligo d(T)$_{25}$ magnetic beads (Dynal product No. 610.02).
3. 2X Binding buffer (2X BB): 20 mM Tris-HCl, pH 7.5, 1.0 M LiCl, 2 mM EDTA if mRNA is isolated from total RNA.
4. Lysis/binding buffer (L/BB): 100 mM Tris-HCl, pH 7.5, 0.5 M LiCl, 10 mM EDTA pH 8, 5 mM dithiotreitol, 1% LiDS, if mRNA is isolated directly from cells or tissue.
5. Washing buffer A: 10 mM Tris-HCl, pH 7.5, 0.15 M LiCl, 1 mM EDTA, 0.1% LiDS, 10 µg/mL mussel glycogen, if mRNA is isolated directly from cells or tissue.

6. Washing buffer B: 10 mM Tris-HCl, pH 7.5, 0.15 M LiCl, 1 mM EDTA, 10 μg/mL mussel glycogen.

   You can also purchase the different buffers with oligo(dT) magnetic beads in two different kits: Dynabeads mRNA purification kit (Dynal No. 610.01) for total RNA, and Dynabeads mRNA DIRECT kit (Dynal No. 610.11) for cells or tissue. (Mussel glycogen is not included in the commercially available buffers.)

   1X First-strand buffer (50 mM Tris-HCl, pH 8.3, 75 mM KCl, 3 mM MgCl$_2$, 10 μg/mL mussel glycogen).

2.2. First- and Second-Strand Synthesis

1. 5X First-strand buffer: 250 mM Tris-HCl, pH 8.3, 375 mM KCl, 15 mM MgCl$_2$.
2. 0.1M Dithiotreitol (DTT).
3. dNTP mix (10 mM each).
4. SUPERSCRIPT™ II Reverse transcriptase (RT) (200 U/μL).
5. RNase OUT™ (40 U/μL).
6. 5X Second-strand buffer: 100 mM Tris-HCl, pH 6.9, 450 mM KCl, 23 mM MgCl$_2$, 0.075 mM β-NAD$^+$, 50 mM ammonium sulfate.
7. *E. coli* DNA ligase (10 U/μL).
8. *E. coli* DNA polymerase (10 U/μL).
9. *E.coli* RNase H (2 U/μL).
   These items are components of the Superscript Choice System for cDNA synthesis (Gibco BRL, cat. no. 18090–019) but can also be purchased separately.
10. Washing buffer C: 5 mM Tris-HCl, pH 7.5, 0.5 mM EDTA, 1 M NaCl, 200 μg/mL SDS, 10μg/mL mussel glycogen.
11. Washing buffer D: 5 mM Tris-HCl, pH 7.5, 0.5 mM EDTA, 1 M NaCl, 1% BSA.
12. EDTA 0.5 M, pH 7.5.
13. 1X Buffer 4: 20 mM Tris-acetate, pH 7.9, 10 mM magnesium acetate, 50 mM potassium acetate, 1 mM DTT, 200 μg/mL BSA.

2.3. Cleavage With Anchoring Enzyme

1. *Nla*III, 10 U/μL (New England Biolabs, cat. no. R0125S): place in aliquots and store at −80°C.
2. 10X buffer 4: 200 mM Tris-acetate, pH 7.9, 100 mM magnesium acetate, 500 mM potassium acetate, 10 mM DTT and 100X BSA.
3. RNase Free LoTE: 3 mM Tris-HCl, pH 7.5, 0.2 mM EDTA, pH 7.5.

2.4. Ligation of T7 SAGE Adapter

1. T4 DNA ligase high concentrate, 5 U/μL (Gibco BRL cat. no. 15224–041).
2. Adapter T7-SAGE (10 ng/μL).
   Order high-quality (PAGE purified) oligonucleotides.
   T7-sage f: 5'cagagaatgcataatacgactcactatagggatccacaagaactactacatg-3'.
   T7-sage r: 5'tagtagttcttgtggatccctatagtgagtcgtattatgcattctctg-3' (with 5' end phosphorylation and C7 amino modification on the 3' end).

Oligos must be resuspended at a 300 μ*M* concentration in STE buffer (10 m*M* Tris-HCl, pH 8, 50 m*M* NaCl, 1 m*M* EDTA).

To anneal the oligonucleotides, mix equimolar amounts of each oligonucleotide and heat at 95°C for 2 min, 65°C for 10 min, 37°C for 10 min and cool down to room temperature for 20 min.

Dilute Adapter T7-SAGE to 10 ng/μL in DEPC water.

2.5. In Vitro Transcription and Recovery of Small Amplified RNA

1. T7 Megascript Kit Ambion (Ambion catalogue number 1334)
2. Microcon YM-10 column (Millipore cat. no. 42406–7)

2.6. Performance Check PCR Reactions

1. Design three oligonucleotides on the same transcript expressed in your tissue(s) of interest. Two forward primers: one upstream of the last CATG site and one downstream of the last CATG site, and one common reverse primer.
2. We choose mouse G3PDH transcript and designed the three following primers:
   -mG3PDH f1: 5' tgacatcaagaaggtggt 3'
   -mG3PDH f2: 5' aaggagtaagaaaccctgga 3'
   -mG3PDH r: 5' cagcgaactttattgatggt 3'
3. Taq polymerase (Qiagen) with dNTP (10 m*M* each) and 10X reaction buffer.

3. Methods

3.1. Isolating mRNA

1. Isolate total RNA with your method of choice. We isolate total RNA from limited amount of microdissected samples with QiaAmp viral RNA mini kit (Qiagen, cat. no.: 52904).
2. You can also directly isolate mRNA from cells or tissue on oligo (dT) magnetic beads (*see* **Note 4.1.**).

3.1.2. Prepare Magnetic Beads

1. Resuspend Dynabeads oligo d(T)$_{25}$, and transfer 20 μL to a new Rnase-free siliconized microcentrifuge tube.
2. Place the tube on magnetic stand for 1 min, and carefully remove the supernatant.
3. To wash the beads, resuspend them in 200 μL 2X binding buffer.
4. Place the tube on magnetic stand for 1 min, and carefully remove the supernatant. Resuspend the beads in 50 μL of 2X binding buffer

3.1.3. Bind mRNA to Magnetic Beads

1. Bring total RNA (*see* **Subheading 3.1.**) to a volume of 50 μL in DEPC treated water.
2. Heat at 65°C for 2 min and mix the RNA with the prewashed beads in a final volume of 100 μL.
3. Incubate at room temperature for 15 to 30 min on a rotor.

4. Of note, if you have a very limited amount of material (few hundred cells) efficiency of mRNA binding to the beads is increased if the binding is performed in a smaller volume (final volume of 20 to 40 μL).

3.1.4. Washing Steps

1. Wash the beads as described (**Note 4.2.**).
2. Wash twice in 200 μL of washing buffer B.
3. Wash four times in 50 μL of 1X transcription buffer.

3.2. First- and Second-Strand Synthesis

1. Prepare transcription mix on ice:
   4 μL of 5X first-strand buffer
   2 μL of 0.1 *M* DTT
   1 μL of dNTP (10 m*M* each)
   0.5 μL of Rnase OUT
   11.5 μL of DEPC treated water
2. After the final wash from **Subheading 3.1.4.**, resuspend the beads in the transcription mix and incubate at 37°C for 2 min.
3. Add 1 μL of Superscript II and slowly vortex.
4. Incubate at 37°C for 1 h, with intermittent gentle mixing.
5. Put on ice and add: (add the enzymes at the end)
   91 μL of ice cold DEPC-water
   30 μL of 5X second-strand buffer
   3 μL of dNTP (10 m*M* each)
   1 μL of *E coli* ligase
   4 μL of *E coli* DNA polymerase
   1 μL of *E coli* Rnase H
   (Final volume of 150 μL)
6. Incubate at 16°C for 2 h, with intermittent gentle mixing.
7. Near the end of this reaction prewarm and place an aliquot of washing buffer C at 75°C.
8. Washing buffer D must be at room temperature (or 37°C). Cold buffer might cause clumping of the beads.
9. Put on ice.
10. Add 10 μL EDTA 0.5 *M*, pH 7.5, to inhibit the reaction, mix well.
11. Place the tube on a magnet for 1–2 min and remove the supernatant.
12. To inactivate *E coli* DNA polymerase, add 150 μL prewarmed washing buffer C and incubate 10 min at 75°C, with intermittent gentle mixing.
13. Place the tube on a magnet for 1–2 min and remove supernatant
14. Wash again with 150 μL of prewarmed washing buffer C.
15. Wash four times with 300 μL of Washing buffer D.
16. Optional: store 15 μL of the resuspended beads for control PCR (*see* **Note 4.4.**).
17. Wash once with 100 μL of 1X buffer 4. Transfer the content of the tube to a new sterile siliconized microcentrifuge tubes (to recover the maximum amount of beads,

if beads attach to microcentrifuge tube, gently scrape them into an extra 50 μL of 1X buffer 4).
18. Wash again with 100 μL of 1X buffer 4.
19. You can store the beads overnight at 4°C, or proceed to the next step.

3.3. Cleavage With Anchoring Enzyme

1. After final wash resuspend the beads in the following mix:
   42.5 μL of LoTE
   0.5 μL of 100X BSA
   5 μL of 10X buffer 4
   2 μL of *Nla* III (10U/μL)
2. Incubate at 37°C for 1 h with intermittent gentle mixing.
3. Prewarm washing buffer C at 65°C, and washing buffer D to room temperature (or 37°C).
4. Wash twice in 200 μL of washing buffer C.
5. Wash four times in 300 μL of washing buffer D.
6. Optional: store 8 μL of the resuspended beads for control PCR (*see* **Note 4.4.**).
7. Wash in 100 μL of 1X ligase buffer. Transfer the content of the tube to a new sterile siliconized microcentrifuge tube.
8. Wash again in 100 μL of 1X ligase buffer.

3.4. Ligation of T7 SAGE Adapter

1. After the last wash resuspend the beads in:
   4.5 μL of LoTE
   3 μL of Adapter T7-SAGE (10 ng/μL)
   2 μL of 5X ligase buffer (be careful, 5X ligase buffer is very viscous).
2. Resuspend the beads by gently flicking the tube with a finger, or slow vortex.
3. Incubate for 2 min at 50°C.
4. Cool the tube to room temperature for 15 min, with intermittent gentle mixing, then chill on ice.
5. Add 0.5 μL of T4 DNA ligase High Concentrate (5 U/μL).
6. Incubate at 16°C for 2 h, with intermittent gentle mixing.
7. Place on ice.
8. Wash four times in 200 μL washing buffer D, prewarmed to room temperature.
9. Wash two to four times in 1X transcription reaction buffer (T7 Megascript Kit, Ambion).

3.5. In Vitro Transcription and Small Amplified RNA Recovery

1. After the last wash resuspend the beads in: (except for enzyme mix, reagents must be at room temperature)
   8 μL of DEPC water
   2 μL of each rNTP (total 8 μL)
   2 μL of 10X reaction buffer
   2 μL of Enzyme mix

2. Incubate for 3 h at 37°C on a rotor or with intermittent gentle mixing.
3. Carefully recover the supernatant.
4. Store the beads in Rnase-free LoTE at 4°C and/or perform additional rounds of amplification.
5. We purify the small amplified RNA on microcon YM-10 column.
6. Wash the column once by adding 50 μL of DEPC water on top of the column, centrifuge for 5 min at 14,000g, discard flow-through.
7. Put the RNA sample on the column in a final volume of 300 μL (complete with DEPC treated water), centrifuge 30 min at room temperature at 14,000g. Discard flow-through. Wash the column with 300 μL DEPC-treated water, centrifuge 30 min at room temperature at 14,000g.
8. Place the column upside down on a new Rnase-free microcentrifuge tube, add 50 μL DEPC treated water and centrifuge 5 min at 1000 to 4500g (make sure you have recovered 40 μL).
9. Purified small amplified RNA is ready for use in a regular SAGE experiment.

4. Notes

1. This protocol was devised for the use of total RNA. We obtained good results with QiaAmp viral RNA mini kit (Qiagen, cat. no: 52904).
   You could also isolate mRNA from cells or tissue directly on magnetic beads. You must use lysis/binding buffer instead of binding buffer to bind mRNA to the beads. In addition, add extra washes with washing buffer A after binding. Refer to manufacturer's instruction for more details.
2. Many different enzymatic steps, and many washing and mixing of the beads are performed throughout the procedure. It is important to loose as few beads as possible. As the beads may stick to the tube it is important that you use no-stick siliconized microcentrifuge tubes. Handle the beads gently, avoid drying, avoid centrifugation. To perform the various washing steps, always:
   a. Put the microcentrifuge tube on a magnetic stand, wait about 1 min until solution is clear.
   b. Carefully remove the supernatant without touching the beads. Be sure that beads do not come with the last drop of washing buffer. If this occurs, pipet back a few μL of the washing buffer, wait for another min and carefully remove last drop of washing buffer.
   c. While the microcentrifuge tube is on the magnetic stand, add the appropriate amount of buffer.
   d. Remove the tube from the magnetic stand and resuspend the beads. To resuspend the beads, gently flick the tube or apply slow speed vortex until the beads are detached from the wall of the tube and the solution is homogeneously brown.
   It is worthwhile noticing that some groups *(16,17)* have experienced better washing without loss of beads by pipeting the beads up and down. Saturate the tip in the buffer you are using for washing step in order to prevent the beads from sticking to the pipet tip, and wash your tip in cleared buffer when the microcentrifuge tube is back on the magnetic stand.

e. Return the microcentrifuge tube to the magnetic stand. Wait 1 min until solution is clear. If some liquid remains in the cap of the microcentrifuge tube, pipet 10 to 20 µL of cleared washing buffer to wash the cap, replace the washing buffer in the microcentrifuge tube and wait 1 min until solution is clear.

f. Go from step a to step f for every wash.

3. Clumping of the beads might occur, especially after second strand synthesis and cleavage with anchoring enzyme.

In order to prevent clumping of the beads:

a. Carefully follow the washing procedure described.

b. Use warm washing buffer C (65°C) and washing buffer D (from room temperature to 37°C), as the SDS present in the washing buffer C might precipitate if cold.

c. Perform washing steps after second strand synthesis as quickly as possible.

d. If clumping occurs, increase the volume of washing buffer, and increase BSA concentration (up to three times) in washing buffer C.

Some protocols *(17)* use 0.1% SDS/200 µg/mL BSA in washing buffer C instead of 1% SDS, they also replace SDS by 1% Tween-20 in washing buffer C for washing steps after cleavage with anchoring enzyme.

4. a. You can check cDNA synthesis and cleavage with anchoring enzyme by PCR.

Set 2 PCR reaction mixes: mix 1 and mix 2

15 µL of 10x reaction buffer

3 µL dNTP (10 m*M* each)

3 µL of mG3PDH f1 primer (10 pmoles/µL) in mix 1 and 3 µL of mG3PDH f2 primer (10 pmoles/µL) in mix 2

3 µL of mG3PDH r primer (10 pmoles/µL)

1 µL *Taq* polymerase

116 µL of water

Prepare PCR reaction as discussed in **Table 1**.

Amplify using the following cycling parameters:

step 1: 95°C 2 min; step 2: 95°C 1 min; step 3: 50°C 1 min; step 4: 72°C 1 min.

Repeat steps 2 t 4 for 28* cycles.

Step 5: 72°C 5 min.

* the number of PCR cycles may have to be optimized, depending on the quantity of your starting material.

Analyze by agarose gel electrophoresis (1,5%):

You should see two 180 and 420 bp PCR bands after second strand synthesis, the 420 bp PCR band (tube 3) should be lost after cleavage with anchoring enzyme.

b. You can check the yield of amplification by PCR.

Retrieve 1% of the resuspended magnetic beads after second strand synthesis before amplification procedure (3 µL from the aliquot from **Subheading 3.2.**) and retrieve 1% of the resuspended magnetic beads after second strand synthesis of the SAGE procedure.

Set PCR reactions:

Add to aliquot of template on magnetic beads

Table 1
The Set Up for Six Different PCR Reactions

	Tube 1	Tube 2	Tube 3	Tube 4	Tube 5	Tube 6
Mix 1	47 μL	—	47 μL	—	47 μL	—
Mix 2	—	47 μL	—	47 μL	—	47 μL
Aliquot from cDNA synthesis	3 μL	3 μL	—	—	—	—
Aliquot from cleavage	—	—	3 μL	3 μL	—	—
Water	—	—	—	—	3 μL	3 μL

  12 μL of 10X reaction buffer
  2.4 μL dNTP (10 mM each)
  2.4 μL of mG3PDH f2 primer (10 pmoles/μL)
  2.4 μL of mG3PDH r primer (10 pmoles/μL)
  0.6 μL *Taq* polymerase
  Add water to a final volume of 120 μL
  Divide in 5 aliquots of 20 μL
  Amplify using the following cycling parameters:
  step 1: 95°C 2 min; step 2: 95°C 1 min; step 3: 50°C 1 min; step 4: 72°C 1 min.
  Repeat steps 2 to 4 for 18, 21, 24, 26, and 28 cycles, respectively. Step 5: 72°C
  5 min.
  (Note that the number of PCR cycles might have to be optimized depending on
  the quantity of your starting material.)
  An example of the yield of amplification is shown in **Fig. 3**.

5. One extra advantage of this procedure for mRNA amplification is that the template is covalently bound to magnetic beads and can thus be reused for successive rounds of amplification in order to increase the yield in a linear way.

   Carefully recover the supernatant after in vitro transcription and resupend the beads in the transcription mix described in **Subheading 3.5.**

   Incubate for 3 h at 37°C on a rotor or with intermittent gentle mixing.

   Repeat as many times as needed. If you store the beads at 4°C, resuspend them in LOTE. Wash the beads in 1X transcription buffer (as described) before subsequent amplification cycles.

6. All synthesized mRNAs and all double-strand cDNAs will have the same 5' extremity. In the SAGE procedure, after cleavage with the anchoring enzyme each cleaved cDNA molecule will generate a 5' gggatccacaagaactactacatg 3', 5' **tagtag ttctt**tgtggatcc 3' double-strand DNA. These molecules, if not properly washed away, will be able to ligate adapters from the SAGE procedure, and be processed as tags. We observed 5% of tags derived from adapter T7-SAGE in our amplified library *(13)*. Some adapter-derived tags will inevitably be generated, therefore, it is impor-

18 21 24 26 28

G3PDH
Amplified
RNA

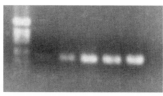

G3PDH non
amplified
RNA

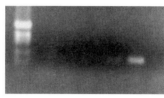

Fig. 3. Evaluation of yield of amplification by PCR. A SAR-SAGE procedure was performed starting from 50 ng total RNA. Same (1%) aliquots were retrieved after second strand synthesis, one during the amplification procedure (**bottom**), one after two rounds of amplification were pooled in a SAGE procedure (**top**). A fragment of G3PDH transcript was amplified for 18, 21, 24, 26, and 28 PCR cycles, using mG3PDH f2 and mG3PDH r primers.

tant that the reverse sequence of the last ten bases of the adapter (sequence bolded) are properly chosen. The tagtagttct tag that we observed with the described T7-SAGE adapter does not correspond to any known transcript in mouse and in human. You may need to modify the sequence of the adapter by replacing the last ten bases before catg if working in other organisms.

7. Small amplified RNA to be used in SAGE procedure
   Once the small amplified RNAs are purified, they can serve as template in SAGE or LongSAGE *(18)* procedure. Several protocols are described for SAGE *(7,17)* (www.sagenet.org/sage-protocol.htm) and a commercial SAGE kit is also available (I-SAGE kit. Invitrogen. Cat. no. T5000-01).

   By slight modification of the adapter (gatc overhang at the 5' end of the reverse primer instead of catg overhang at the 3' end of the sens primer), this procedure could theoretically be used upstream of the SADE procedure (please refer to the SADE technique for correct construction of primers *[6,16]*).

8. In order to limit possible bias we have opted for a procedure that does not rely on additional PCR steps to generate SAGE libraries from small amount of input RNA. The linear character of the amplification by T7-dependent transcription introduces minimal biases in the generation of cDNA libraries or in the preparation of labeled targets for microarray experiments *(19–21)*.

   Following this protocol, we were able to generate a SAGE library from 50 ng total RNA derived from adult mouse thyroids.

The expression profile was consistent with what is known of thyroid metabolism and showed similitude with a described human thyroid SAGE library *(22)*.

We compared the results obtained with this protocol from 50 ng total RNA to those of a library obtained from 5 µg of the same RNA preparation with the SAGE protocol *(7)*.

In order to validate our results *(13)* we established the Pearson correlation between the two libraries after a variance stabilizing transformation. The observed correlation of 75% compares well with microarray experiments *(23)* in which results obtained with PCR or linear amplification are compared with those from nonamplified samples (**Fig. 4**). However, when individual tags were analyzed, significant discrepancies were observed, with under representation of some relatively frequent tags found in the amplified library. These differences were observed for 4.5% of the most abundant tags. They are similar to those described previously for linear amplification in microarray experiments *(19)*.

In conclusion, the SAR-SAGE technique offers an interesting alternative to pre-PCR amplification in the preparation of SAGE libraries from small amounts of (micro-dissected) tissue samples.

Some of the troubleshootings are described in **Table 2**.

Acknowledgments

C.Vilain is "Research Fellow" of the Fonds National de la Recherche Scientifique. We thank Sabine Costagliola, Frederick Libert and Jean-Marc Elalouf for constructive discussions. Supported by the Belgian State, Prime Minister's office, Service for Sciences, Technology and Culture, grants from the FRSM, FNRS and Association Recherche Biomédicale et Diagnostic.

References

1. Velculescu, V. E., Zhang, L., Vogelstein, B., and Kinzler, K. W. (1995) Serial analysis of gene expression. *Science* **270**, 484–487.
2. Polyak, K. and Riggins, G. J. (2001) Gene discovery using the serial analysis of gene expression technique: implications for cancer research. *J. Clin. Oncol.* **19**, 2948–2958.
3. Logan, M. (2002) SAGE profiling of the forelimb and hindlimb. *Genome Biol.* **3**, REVIEWS1007.
4. Munasinghe, A., Patankar, S., Cook, B. P., et al. (2001) Serial analysis of gene expression (SAGE) in *Plasmodium falciparum*: application of the technique to A-T rich genomes. *Mol. Biochem. Parasitol.* **113**, 23–34.
5. Chen, J., Sun, M., Lee, S., Zhou, G., Rowley, J. D., and Wang, S. M. (2002) Identifying novel transcripts and novel genes in the human genome by using novel SAGE tags. *Proc. Natl. Acad. Sci. USA* **99**, 12257–12262.
6. Virlon, B., Cheval, L., Buhler, J. M., Billon, E., Doucet, A., and Elalouf, J. M. (1999) Serial microanalysis of renal transcriptomes. *Proc. Natl. Acad. Sci. USA* **96**, 15286–15291.

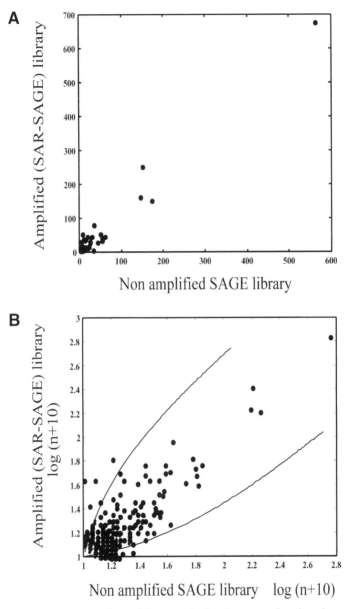

Fig. 4. Graphic representation of the correlation between the abundance of tags in two libraries (a SAR-SAGE library from 50 ng total RNA, and a SAGE library from 5 μg of the same RNA preparation): **A.** using the raw values of tag abundance per 10,000 tags in direct plot, **B.** after variance stabilizing transformation ($\log 10(n+k)$, where n is the absolute number of tag in the library (normalized for 10,000 tags) and k is a constant which was set to 10. The two lines delineate a confidence interval for tags showing less than fivefold variation.

Table 2
Troubleshooting

Problem	Cause	Solution
No bands seen after cDNA synthesis when analysed by PCR	Poor quality RNA, RNA degraded	It is difficult to assess RNA quality when working with very limited amount of material Use extreme care while handling RNA samples to prevent RNase contamination Check quality of RNA by RT-PCR for transcripts known to be expressed in your tissue of interest
	Poor binding to magnetic beads	When very limited amount of material is used, binding in conventional volume might be poor. Perform binding in a volume as low as possible (20–40 μL)
	cDNA synthesis reagent not working	Perform cDNA synthesis with control RNA of known quality
PCR band for longer transcript does not disappear after cleavage	Number of PCR cycles too high	Make semi-quantitative PCR as described in 4,4,a to ensure that substantial amount of cDNA is cleaved
	NlaIII not working well	Cleave a plasmid vector to asses efficiency of enzyme
Clumping of the beads	Precipitation of SDS in washing buffer C, and entangling of doubled stranded cDNA	Keep solution warm to prevent SDS precipitation Perform washing steps quickly Increase washing volumes Increase number of washes Modify washing buffer as described in section 4,3: decrease SDS concentration to 0,1%, and add 200 μg/mL BSA in buffer C

Inefficient T7-sage adapter ligation	CATG overhang may be cleaved by exonuclease activity of residual E coli polymerase	Heat in wash buffer C to remove exonuclease activity, as described Thoroughly wash the beads after second strand synthesis
	Adapter incorrect	Check for correct annealing by gel electrophoresis Check for phosphorylation by self ligation and gel electrophoresis
Overall poor performance and poor RNA amplification yield	Improper mixing and washing of the beads	Be sure to thoroughly resuspend the beads after every washing steps and to mix the beads intermittently during enzymatic reactions
	Loss of beads	Use siliconized microcentrifuge tubes When pipetting, saturate the tip in washing buffer and rince your tip in buffer Scrape the sides of the tube to remove any beads that stick to the tubes especially when changing tube
	T7 megascript not working	Use pTRI-Xef control following manufacturer's instructions

7. St Croix, B., Rago, C., Velculescu, V., et al. (2000) Genes expressed in human tumor endothelium. *Science* **289**, 1197–1202.
8. Ye, S. Q., Zhang, L. Q., Zheng, F., Virgil, D., and Kwiterovich, P. O. (2000) miniSAGE: gene expression profiling using serial analysis of gene expression from 1 microg total RNA. *Anal. Biochem.* **287**, 144–152.
9. Peters, D. G., Kassam, A. B., Yonas, H., et al. (1999) Comprehensive transcript analysis in small quantities of mRNA by SAGE-lite. *Nucleic Acids Res.* **27**, e39.
10. Datson, N. A., Perk-de Jong, J., van den Berg, M. P., de Kloet, E. R., and Vreugdenhil, E. (1999) MicroSAGE: a modified procedure for serial analysis of gene expression in limited amounts of tissue. *Nucleic Acids Res.* **27**, 1300–1307.
11. Neilson, L., Andalibi, A., Kang, D., et al. (2000) Molecular phenotype of the human oocyte by PCR-SAGE. *Genomics* **63**, 13–24.
12. Lee, S., Chen, J., Zhou, G., and Wang, S. M. (2001) Generation of high-quantity and quality tag/ditag cDNAs for SAGE analysis. *Biotechniques* **31**, 348–354.
13. Vilain, C., Libert, F., Venet, D., Costagliola, S., and Vassart, G. (2003) Small Amplified RNA-SAGE: an alternative approach to study transcriptome from limiting amount of mRNA. *Nucleic Acids Res.* **31**, e24.
14. Eberwine, J., Yeh, H., Miyashiro, K., et al. (1992) Analysis of gene expression in single live neurons. *Proc. Natl. Acad. Sci. USA* **89**, 3010–3014.
15. Marble, H. A. and Davis, R. H. (1995) RNA transcription from immobilized DNA templates. *Biotechnol. Prog.* **11**, 393–396.
16. Cheval, L., Virlon, B., and Elalouf, J. M. (2000) SADE: a microassay for serial analysis of gene expression, in *Functional genomics: a practical approach* (Hunt, S. P. and Liversey, J. P., ed.), Oxford University press, 139–163.
17. Blackshaw, S., Kim, J. B., St Croix, B., and Polyak, K. (2003) Serial Analysis of Gene Expression. In: Current Protocols in Molecular Biology (Ausubel, F. M., Brent, R., Kingston, R., et al., ed.) Massachusett's General Hospital, Harvard Medical School, 25B.6.1–25B.6.29
18. Saha, S., Sparks, A. B., Rago, C., et al. (2002) Using the transcriptome to annotate the genome. *Nat. Biotechnol.* **20**, 508–512.
19. Wang, E., Miller, L. D., Ohnmacht, G. A., Liu, E. T., and Marincola, F. M. (2000) High-fidelity mRNA amplification for gene profiling. *Nat. Biotechnol.* **18**, 457–459.
20. Van Gelder, R. N., von Zastrow, M. E., Yool, A., Dement, W. C., Barchas, J. D., and Eberwine, J. H. (1990) Amplified RNA synthesized from limited quantities of heterogeneous cDNA. *Proc. Natl. Acad. Sci. USA* **87**, 1663–1667.
21. Baugh, L. R., Hill, A. A., Brown, E. L., and Hunter, C. P. (2001) Quantitative analysis of mRNA amplification by in vitro transcription. *Nucleic Acids Res.* **29**, e29.
22. Pauws, E., Moreno, J. C., Tijssen, M., Baas, F., de Vijlder, J. J., and Ris-Stalpers, C. (2000) Serial analysis of gene expression as a tool to assess the human thyroid expression profile and to identify novel thyroidal genes. *J. Clin. Endocrinol. Metab.* **85**, 1923–1927.
23. Iscove, N. N., Barbara, M., Gu, M., Gibson, M., Modi, C., and Winegarden, N. (2002) Representation is faithfully preserved in global cDNA amplified exponentially from sub-picogram quantities of mRNA. *Nat. Biotechnol.* **20**, 940–943.

10

Gene Expression Informatics

Martin Leach

Summary
There are many methodologies for performing gene expression profiling on transcripts, and through their use scientists have been generating vast amounts of experimental data. Turning the raw experimental data into meaningful biological observation requires a number of processing steps; to remove noise, to identify the "true" expression value, normalize the data, compare it to reference data, and to extract patterns, or obtain insight into the underlying biology of the samples being measured. In this chapter we give a brief overview of how the raw data is processed, provide details on several data-mining methods, and discuss the future direction of expression informatics.

Key Words: Bioinformatics, clustering, data analysis, databases, gene-expression, microarrays, software

1. Introduction

On April 14, 2003 the International Human Genome Sequencing Consortium, led in the United States by the National Human Genome Research Institute (NHGRI) and the Department of Energy (DOE), announced the successful completion of the Human Genome Project *(1)*. Now, researchers for the first time have the complete set of data for studying gene makeup and understanding gene regulation. However, there is still an active debate as to how many genes are actually in the human genome. Initial publications based on a draft of the human genome cited between 24,500 *(1)* and 26,383 *(2)* genes. This is approximately half of the mean "estimate" in the Gene Sweepstake (http://www.ensembl.org/Genesweep) but the number was verified as being at least 24,500 genes at the 68th Cold Spring Harbor Symposium on Quantitative Biology. The definitive number of genes will remain unknown for a number of years until

From: *Methods in Molecular Biology, Vol. 258: Gene Expression Profiling: Methods and Protocols*
Edited by: R. A. Shimkets © Humana Press Inc., Totowa, NJ

millions of proprietary expressed sequence tag (EST) sequences from companies such as Incyte Pharmaceuticals, Human Genome Sciences, Millenium Pharmaceuticals, and CuraGen Corporations are combined with the public data.

There are many molecular biology techniques for the capture and measurement of gene transcripts, many of which are presented in this book. Before utilizing microarray or other expression measurement technologies some thought needs to be applied to proper experiment design so that statistically significant observations can be generated. Kerr et al. *(3)* gives a good overview of how researchers should approach experimental design as it pertains to expression profiling. Unfortunately, researchers spend a disproportionate amount of time in experimental design in the rush to examine expression data. This approach typically results in a qualitative measure of expression levels and the data is tossed over the proverbial fence to the informatics scientists to identify patterns and give clarity using computational techniques. The desire to extract meaningful results and an understanding of gene regulation and association of gene expression levels to a desired pathophysiological state has resulted in a plethora of techniques and software that is bewildering to researchers. Lorkowski et al. *(4)* presents an excellent review of computation methods, and bioinformatics tools are presented as well. In this chapter , the basics of expression profiling analysis and analytical methods will be presented, focusing predominantly on microarray expression analysis .

2. Materials

2.1. Expression Data Sources

One of the most comprehensive reference collections of gene expression microarray data can be found at Gene Expression Omnibus (http://www.ncbi. nlm.nih.gov/geo/) and is maintained by the National Center for Biotechnology Information (NCBI) *(5)*. The data is comprised of noncommercial, commercial, or custom nucleotide microarrays with some transcript expression available from serial analysis of gene expression (SAGE) experiments *(6)*. A central site for collecting and organizing SAGE data on cancer tissues can be found at SAGENET (http://www.sagenet.org/resources/data.htm). A larger set of oncology SAGE data can be found at The Cancer Genome Anatomy Project (CGAP). Fortunately, the CGAP SAGE data has been deposited into the NCBI GEO database. A group in France has adapted the SAGE methodogy with a SAGE Adaptation for Downsized Extracts (SADE) *(7–8)* and has provided data (http://www-dsv.cea.fr/thema/get/sade.html). **Table 1** lists the predominant sources of expression data for a variety of organisms where scientists can download, manipulate, and perform further data-mining experimentation. We have found that these data repositories are useful because in combination with our own

Table 1
Frequently Used Gene Expression Data Repositories

Description	URL	Comment
ArrayExpress	http://www.ebi.ac.uk/arrayexpress	Public repository for microarray data in accordance with HGED standards.
BodyMap	http://bodymap.ims.u-tokyo.ac.jp/	Human and mouse gene expression database using ESTs.
Brown Lab, Stanford	http://cmgm.stanford.edu/pbrown/explore/	Searchable database of published yeast microarray data.
ExpressDB	http://arep.med.harvard.edu/ExpressDB/	Yeast and E. coli RNA expression data.
Gene Expression Omnibus	http://www.ncbi.nlm.nih.gov/geo/	Compendium of expression data from many platforms for several organisms.
HuGE	http://zlab.bu.edu/HugeSearch/	A database of human gene expression using arrays.
Jackson Labs.	http://www.jax.org/staff/churchill/labsite/datasets/index.html http://www.informatics.jax.org/	Many mouse microarray datasets.
SAGENET	http://www.sagenet.org/resources/index.html	SAGE data available for download from many cancer tissues samples.
Stanford Microarray Database	http://www.dnachip.org/	>40,500 microarray experiments covering 25 organisms
Yeast Expression Data	http://web.wi.mit.edu/young/expression/	Genome-wide expression data and detailed information on yeast mRNAs.

experimental data they have provided confirmatory evidence to our initial discoveries (unpublished).

2.2. Informatics Software

There are many commercial, academic, and freely available platforms or applications for gene expression analysis. Software that is most widely used includes Rosetta Resolver (Rosetta Inpharmatics, Kirkland, Washington), GeneSpring™ (Silicon Genetics, Redwood City, CA), S-Plus® (Insightful Corporation, Seattle, Washington), MatLab® (The Mathworks Inc., Natick, MA), and Spotfire DecisionSite (Spotfire Inc., Boston, MA). However, these software applications and data warehouses are all commercial. Academic researchers performing detailed expression analysis and modeling have generated many software applications and web-based interfaces (*see* **Table 2**). One platform that requires mention is The R Project for Statistical Computing (http://www.r-project.org). Similar to commercial software applications such as MatLab, the R Project provides a comprehensive framework for performing powerful statistical analyses and data visualization. Furthermore, many modules and packages have been developed specifically for expression data processing, analysis, and visualization (http://www. stat.uni-muenchen.de/~strimmer/rexpress.html). **Table 2** contains a list of the most popular software applications or frameworks that are available for use in expression profiling data analysis and visualization.

3. Methods

3.1. Raw Data Handling

Recent technologies for gene expression analysis have made it possible to simultaneously monitor the expression pattern of thousands of genes. Therefore, all differentially expressed genes between different states (e.g., normal vs diseased tissue) can be easily identified, leading to the discovery of diseased genes or drug targets. One difficulty in identifying differentially expressed genes is that experimental measurements of expression levels include variation resulting from noise, systematic error, and biological variation. Distinguishing the true from false differences has presented a challenge for gene expression analysis. One of the major sources of noise in gene expression experiments is the difference in the amount or quality of either mRNA or cDNA biological material analyzed, or the analytical instruments performing the measurement of gene expression.

In order to address these and other difficulties, methodologies are applied for normalizing, scaling, and difference finding for gene expression data. These methods are applicable to most expression profiling methods but differ according to the idiosyncrasies of each technology. A typical approach to "cleaning" or "processing" the raw data and using it for differential gene analysis is as follows:

- Define noise
- Perform normalization
- Adjust data through scaling
- Compare data from experiments to identify differences
- Perform analytical and data-mining analyses

As researchers are publishing large sets of expression data and they are reused or recombined with other experiments it is important that normalization and other data transformations are described in detail with the publication. To facilitate the sharing of expression data and standardization of microarray expression data sets a Normalization Working Group of the Microarray Gene Expression Data (MGED) organization (http://www.mged.org) has been formed and is attempting to define standards through participation of the scientific community. In addition, it is now required that manuscripts submitted to the journal *Nature* have corresponding microarray data submitted to the GEO or ArrayExpress databases.

3.1.1. Defining Noise

Typically with microarray methods a predominant source of noise results from electrical noise from the microarray scanner. This results in noise values varying between scanners. A simple method used for setting the noise baseline is to determine the average intensity of a low percentage of the signals generated in an expression profiling experiment. For microarray experiments, this is a relatively simple process, for example, the bottom 2% of signals may be collected to generate the noise baseline (9). However, the process of identifying the bottom percentage of low signals is a difficult process in differential display techniques where multiple peaks of intensity are generated for multiple genes in a single electrophoretic data stream (10). Once the noise baseline has been generated it is simply extracted from the experimental measurements to determine the measured value.

3.1.2. Normalization

Normalization is the method of reducing sample-to-sample, batch-to-batch, or experiment-to-experiment variation. A more detailed discussion on the sources of variation can be found in Hartemink et al. (11). Internal standards not expected to change are used for normalization. Multiple housekeeping genes have been identified and a combination of these should be used for normalization purposes (12). It is wise to monitor and periodically evaluate potential changes in gene expression with the housekeeping genes as they are subject to gene regulation.

Table 2
List of Commonly Used Public and Proprietary Expression Analysis and Visualization Software

Description	URL	Comment
ArrayDB 2.0	http://research.nhgri.nih.gov/arraydb/	A software suite that provides an interactive user interface for the mining and analysis of microarray gene expression data.
Array Designer Software	http://www.arrayit.com	Commercial
Bioconductor	http://www.bioconductor.org	Collaborative open-source project to develop a modular framework for analysis of genomics data. Contains modules for microarray analysis.
BioDiscovery	http://www.biodiscovery.com/imagene.asp	BioDiscoverys ImaGene Image Analysis Software
BioSap (Blast Integrated Oligo. Selection Accelerator Package)	http://biosap.sourceforge.net	Public—Oligo design and analysis software for microarrays.
DEODAS (Degenerate Oligo Nucleotide Design and Analysis System)	http://deodas.sourceforge.net/	Public—Oligo design and analysis software for microarrays.
ExpressYourself	http://bioinfo.mbb.yale.edu/expressyourself/	Public—Automated platform for signal correction, normalization, and analyses of multiple microarray format.
Featurama	http://probepicker.sourceforge.net	Public—Oligo design and analysis software for microarrays.
GeneChip® LIMS data warehouse	http://www.affymetrix.com/products/software/index.affx	Commercial Affymetrix Software for GeneChip microarray design and analysis.
GeneSpring™	http://www.sigentics.com/	Commercial software and data warehouse.
GeneX-lite	http://www.ncgr.org/genex	Freely available system for microarray analysis built on open-source software.

Name	URL	Description
GenoMax Gene Expression Analysis Module	http://www.informaxinc.com	Commercial software and data warehouse.
OligoArray	http://berry.engin.umich.edu/oligoarray	Genome-scale oligo design software for microarrays.
Oligos4Array	http://www.mwg-biotech.com/html/d_diagnosis/d_software_oligos4array.shtml	Commercial automated high throughput oligo design software.
R project	http://www.r-project.org (see below for expression analysis modules to use with R)	The R system is a free (GNU GPL) general purpose computational environment for the statistical analysis of data (*33*).
R packages for expression analysis	http://www.stat.uni-muenchen.de/~strimmer/rexpress.html	Many R packages (modules) developed to analyze gene expression from multiple expression profile platforms.
Rosetta Resolver	http://www.rosettabio.com/products/resolver/default.htm	Commercial software and data warehouse.
SAGE analysis software	http://cgap.nci.nih.gov/SAGE http://www.sagenet.org/resources/index.html http://www.ncbi.nlm.nih.gov/SAGE/	
ScanAlyze	http://rana.lbl.gov/EisenSoftware.htm	Michael Eisen's software for processing images from microarrays, and performing multiple forms of data analysis.
SNOMAD	http://pevsnerlab.kennedykrieger.org/snomadinput.html	Web-based software for standardization and normalization of microarray data.
Spotfire DecisionSite	http://www.spotfire.com/products/decision.asp	Commercial visualization and analysis software.

3.1.3. Scaling

Scaling is the process of transforming the expression data points through the application of a scaling factor. This is performed when experimental replicates have been generated, and to facilitate later comparison with a reference set of data. They are scaled so median intensities are the same across the replicates *(13)*. The choice of a local vs global scaling method is important and is dependent on the gene expression changes occurring in the experiment. If the majority of transcripts will be exhibiting an expression change, then a global scaling method should be applied. Alternatively, if a small number of expression changes are expected, then a local (or selected) scaling method should be applied. Often, however, an overall scaling is not sufficient to discriminate between true differences and those attributed to noise. One source of difficulty is identifying the particular genes for use as scaling landmarks. In addition, a nonuniform tapering of the signal across the set of measurements may generate additional noise. The best method of scaling for any given technology should be empirically determined after the use of many replicates of standard samples.

It is beyond the scope of this chapter to discuss outlier detection and we refer readers to Li et al. *(14)* for a detailed discussion.

3.2. Differential Analysis

The purpose of many expression profiling experiments is for the comparison of gene expression levels between two or more states (e.g., diseased vs normal, different time points, or drug treatments and concentrations). One approach for the comparison is to generate a ratio of expression level from state I to state II. For example, state I = 400 U, state II = 200 U, expression ratio = 400/200 or the expression level in state I is twofold higher than that of state II. However, a twofold decrease in gene expression from state I to state II would be represented by 0.5. The result of this simple ratio is a numerical value that has a different magnitude for the same relative effect. An alternative approach that properly handles the magnitude of change is to use a logarithm base 2 transformation *(13)*. By following the examples: $\log_2(100/100) = 0$, $\log_2(200/100) = 1$, $\log_2(100/200) = -1$, $\log_2(400/100) = 2$, $\log_2(100/400) = -2$, we see a symmetric treatment of expression ratios through this logarithmic transformation. This results in an easier interpretation of expression differences.

3.3. Analytical and Data-Mining Analyses

Where large data sets are generated, a number of different algorithms and methods can be applied for the mining and extraction of meaningful data. A common approach is to use cluster analyses to group genes with a similar pattern of gene expression *(15)*. Clustering methods can be divided into two classes, unsupervised and supervised *(16)*. In supervised clustering, distances are created through measurements based on the expression profiles and annotations, whereas unsupervised clustering is based on the measurements themselves. Before expression results from samples can be clustered, a measure of distance must be generated between the observations. A number of different measurements can be used to measure the similarity between any two genes, such as, Euclidean distance or the use of a standard correlation coefficient. A detailed description of distance measurements can be found in Hartigan et al *(17)*. The Euclidean distance measure goes back to simple geometry where the two points "A" and "B" are mapped using (x, y) coordinates in two-dimensional space and a right-angle triangle is formed. The hypotenuse represents the distance between the two points and is calculated using the Pythagorean formula. The (x, y) coordinates for any given point may represent the gene expression level in two states or expression level in one state and another measurement or annotation on the gene. Hence, a drawback of simple distance measurement technique, such as the Euclidean distance is that it allows only expression value and a single state to be compared.

Through the creation of "distances" between any given data point, one-dimensional, two-dimensional, or multidimensional analyses can be generated for numerous genes across multiple states *(17)*. The result is a grouping of similar expression patterns for the different genes. The interpretation is that the similarly clustered or grouped genes are being regulated through a common gene regulation network or pathway. Two-dimensional analyses allow a better dissection of the gene expression pattern as the scientists can manually or automatically subgroup based on physiological or clinical properties of the experimental samples or annotations on the genes being measured *(18)*. Clustering analysis is also used to perform "guilt-by-association" type experiments where the function of an unknown gene is inferred by it's apparent clustering with a gene of known function. This is typically performed using an unsupervised clustering method and has been applied on a large scale with model organisms such as *Saccharomyces cerevisiae (15,19)*.

There are many methods of clustering algorithm. Common methods include: K-mean clustering algorithms *(4,20)*, hierarchical algorithms *(15,18)*, and Self-Organizing-Maps (SOMs) *(16,21)*.

With hierarchical clustering, a dissimilarity measure is created between data points, clusters are formed, and a dissimilarity measurement is created between the clusters. Clusters are merged, distance is recalculated, clusters are broken or merged, and the process is repeated until there is one cluster containing all data points with distances between each data point. A drawback of the hierarchical clustering method is that it is computationally and memory intensive and gives poor performance on large datasets. In addition, when data points are falsely joined in a cluster it is difficult to computationally resolve such problems resulting in a spurious hierarchical organization.

The K-means approach is a much faster clustering method that is more suitable for large-scale applications. K-means is a partitioning method where there are "K" randomly generated "seed" clusters. Each of the data points are associated with each of the clusters based on similarity and the mean of each cluster is generated. The distance from each K-mean is calculated using a Euclidean distance measurement, clusters are reconfigured based on distances, and the process is repeated until there is no significant change to the clusters. One problem of this approach is that the technique cannot adequately deal with overlapping clusters and has a problem with outliers.

The above methods are applicable to datasets when simple one or two-dimensional clustering is required. Multidimensional datasets can be analyzed with complex techniques such as Principal Component Analysis (17,22–23). Principal Component Analysis reduces the complexity of the data by transforming the dataset into variables (eigenvectors and eigenvalues) that are then eliminated once their contribution is assessed. The variables are eliminated in a way so as little loss of information as possible. Each variable is assessed to see how it contributes to the overall variance in the experimental comparison. Values and variables that contribute little variance are removed resulting in a minimalization and identification of values and data dimension(s) that cause the observed effect. Similarly, analysis of variance analysis (ANOVA) and variations on ANOVA can be applied to less complicated datasets (24).

A method used recently for data-mining purposes is support vector machines (SVMs). SVMs are a supervised computer learning method that utilizes known information about expressed genes through the construction of a training set of data. The trained SVM is then applied to unknown genes to identify similarities. There are several forms of SVM techniques, common forms include the Fisher's linear discriminant (25), Parzen windows (26), and decision-tree learning (27). SVMs have several advantages over hierarchical clustering and SOMs in that they have the ability to generate distance matrices in a multidimensional space, they can handle large datasets and identity outliers (28).

The above listing of techniques represents only the commonly used methods. For a more detailed description of clustering and analysis methodologies see Eisen et al. (15), Alter et al. (22), Wu et al. (19), and Lorkowski et al. (4).

3.4. Summary and Future of Expression Informatics

A good deal of work has been performed on the design, processing, and analysis of expression data. A recent trend in genomics and proteomics has been to understand the complex interactions between proteins and genes through signal transduction pathways and regulatory networks. This has been referred to as Systems Biology. Computer scientists have attempted to map out the dynamic behavior of gene expression pathways to map them to a networked architecture. The generation of genetic networks attempts to completely reverse engineer the underlying regulatory interactions using Boolean *(29)* and Bayesian Networks *(30)*. Systems Biology, or as most biologists call it, "Physiology" is complex, and interactions occur across vast temporal and spatial distances in a whole organism. Fully mapping out physiological processes work is needed to integrate the many disparate types of biological data and map them to expression data. Furthermore, modeling a system in it's entirety will be a computationally expensive process that requires vast amount of computational power. Fortunately, as projects such as IBM's Blue Gene mature they may result in a solution for handling these vast computational problems.

Finally, having the core set of genes is useful and will be a powerful resource for scientists, however, the ideal resource for researchers studying gene expression is to have the comprehensive database of gene variants. Gene variants can be broken down into two major categories; variants that are consistent within individuals brought about through alternate splicing of the gene primary transcript, or variants that result from genotypic differences between individuals in a given population. Recent technological advantages with the creation of high-density microarrays have allowed scientists to perform gene expression analysis at the genome scale *(31–32)*. Following deposition of closely mapped genomic fragments or gene candidates, subsequent profiling across multiple biological samples has allowed scientists to perform gene and splice variant identification with significant success *(9)*.

Understanding the precise control of splice variants and their association with specific physiological or pathological states is the ultimate goal of gene expression studies. We are still several years from effectively doing this as the human transcriptome has yet to be fully mapped out and splice variants to be fully identified. As we learn more about the transcriptome, and as technology and data analysis methods advance, we will be able to perform gene variant expression with greater specificity. With this specificity, we will accurately be able to map gene variants to biological systems so that we can simulate them.

References

1. International Human Genome Sequencing Consortium. Initial sequencing and analysis of the human genome. (2001) *Nature* **409,** 860–921.
2. Venter, J. C., Adams, M. D., Myers, E. W., et al. (2002) The sequence of the Human Genome. *Science* **291,** 1304–1351.
3. Kerr, M. K. and Churchill, G. A. (2001) Experimental design for gene expression microarrays. *Biostatistics* **2,** 183–201.

4. Lorkowski, S. and Cullen, P. (eds.) (2003) Computational methods and bioinformatics tools, in *Analysing Gene Expression: A handbook of methods possibilities and pitfalls* ,Wiley-VCH, Weinheim, Germany, 769–904.

5. Edgar, R., Domrachev, M., and Lash, A. E. (2002) Gene Expression Omnibus: NCBI gene expression hybridization array data repository. *Nucleic Acids Res.* **1,** 207–210.

6. Velculescu, V., Zhang, L., Vogelstein, B., and Kinzler, K. (1995) Serial analysis of gene expression. *Science* **270,** 484–487.

7. Cheval, L., Virlon, B., and Elalouf, J. M. (2000) SADE: a microassay for serial analysis of gene expression, in *Functional Genomics: a practical approach* (Hunt, S. and Livesey, J., eds.), Oxford University Press, New York, NY, 139–163.

8. Virlon, B., Cheval, L., Buhler, J. M., Billon, E., Doucet, A., and Elalouf, J. M. (1999) Serial microanalysis of renal transcriptomes *Proc. Natl. Acad. Sci. USA* **26,** 15286–15291.

9. Hu, G. H., Madore, S. J., Moldover, B., et al. (2001) Predicting splice variant from DNA chip expression data. *Genome Res.* **7,** 1237–1245.

10. Shimkets, R. A., Lowe, D. G., Tai, J. T., et al. (1999) Gene expression analysis by transcript profiling coupled to a gene database query. *Nat. Biotechnol.* **8,** 798–803.

11. Hartemink, A. J., Gifford, D. K., Jaakkola, T. S., and Young, R. A. (2001) Maximum-likelihood estimation of optimal scaling factors for expression array normalization, in *Microarrays: Optical Technologies and Informatics* (Bittner, M. L., Yidong, C., Dorsel, A. N., and Dougherty, E. R., ed.) SPIE—The International Society for Optical Engineering, Bellingham, WA 132–140.

12. Warrington, J. A., Nair, A., Mahadevappa, M., and Tsyganskaya, M. (2001) Comparison of human adult and fetal expression and identification of 535 housekeeping/maintenance genes. *Physiol. Genomics* **3,** 143–147.

13. Quackenbush, J. (2002) Microarray data normalization and transformation. *Nat. Genet. Suppl.* **32,** 496–501.

14. Li, C. and Wong, W. H. (2001) Model-based analysis of oligonucleotide arrays: Expression index computation and outlier detection. *Proc. Natl. Acad. Sci. USA* **1,** 31–36.

15. Eisen, M. B., Spellman, P. T., Brown, P. O., and Botstein, D. (1998) Cluster analysis and display of genome-wide expression patterns. *Proc. Natl. Acad. Sci. USA* **95,** 14863–14868.

16. Kohonen, T., Huang, T. S., and Schroeder, M. R. (eds.) *Self-Organizing Maps.* Springer-Verlag, New York, NY.

17. Hartigan, J. (ed.) (1975) *Clustering Algorithms* John Wiley and Sons, New York, NY.

18. Alon, U., Barkai, N., Notterman, D. A., et al. (1999) Broad patterns of gene expression revealed by clustering analysis of tumor and normal colon tissues probed by oligonucleotide arrays. *Proc. Natl. Acad. Sci. USA* **12,** 6745–6750.

19. Wu, L. F., Hughes, T. R., Davierwala, A. P., et al. (2002) Large-scale prediction of *Saccharomyces cerevisiae* gene function using overlapping transcriptional clusters. *Nat. Genet.* **31,** 255–265.

20. Tavazoie, S., Hughes, J. D., Campbell, M. J., Cho, R. J., and Church, G. M. (1999) Systematic determination of genetic network architecture. *Nat. Genet.* **22,** 281–285.

21. Tamayo, P., Slonim, D., Mesirov, J., et al. (1999) Interpreting patterns of gene expression with self-organizing maps: Methods and application to hematopoietic differentiation. *Proc. Natl. Acad. Sci. USA* **96,** 2907–2912.

22. Alter, O., Brown, P. O., and Botstein, D. (2000) Singular value decomposition for genome-wide expression data processing and modeling. *Proc. Natl. Acad. Sci. USA* **18,** 10101–10106.

23. Yeung, K. Y. and Ruzzo, W. L. (2001) Principal component analysis for clustering gene expression data. *Bioinformatics* **9,** 764–774.

24. Kerr, M. K., Martin, M., and Churchill, G. A. (2000) Analysis of variance for gene expression microarray data. *J. Comp. Biol.* **7,** 819–837.

25. Duda, R. O. and Hart, P. E. (eds.) (1973) *Pattern Classification and Scene Analysis.* John Wiley and Sons, New York, NY.

26. Bishop, C. (ed.) (1995) *Neural Networks for Pattern Recognition.* Oxford University Press, New York, NY.

27. Quinlan, J. R. (ed.) (1997) C4:5: *Programs for Machine Learning.* Morgan Kaufmann, San Francisco, CA.

28. Brown, M. P., Grundy, W. N., Lin, D., et al. (2000) Knowledge-based analysis of microarray gene expression data by using support vector machines. *Proc. Natl. Acad. Sci. USA* **1,** 262–267.

29. D'haeseleer, P., Liang, S., and Somogyi, R. (2000) Genetic network inference: from co-expression clustering to reverse engineering. *Bioinformatics* **8,** 707–726.

30. Hartemink, A. J., Gifford, D. K., Jaakkola, T. S., and Young, R. A. (2001) Using graphical models and genomic expression data to statistically validate models of genetic regulatory networks. *Pac. Symp. Biocomput.* **6,** 422–433.

31. Chee, M., Yang, R., Hubbell, E., et al. (1996) Accessing genetic information with high-density DNA arrays. *Science* **274,** 610–614.

32. Lipshutz, R. J., Fodor, S. P. A., Gingeras, T. R., and Lockhart, D. J. (1999) High density synthetic oligonucleotide arrays. *Nat. Genet. Suppl.* **21,** 20–24.

33. Ihaka, R. and Gentleman, R. (1996) R: A language for data analysis and graphics. *J. Comput. Graph. Stat.* **3,** 299–314.

Index